SUPPRESSORS

RECOIL Magazine's Complete Guide to Buying, Maintaining, and Shooting with a Silencer

COMPILED BY
THE EDITORS OF
RECOIL MAGAZINE

Published by

RECOIL

Recoil Books, an imprint of Caribou Media Group, LLC
5600 W. Grande Market Drive, Suite 100
Appleton, WI 54913
recoilweb.com | gundigest.com

To order books or other products call 920.471.4522 ext.104
or visit us online at gundigeststore.com

ISBN-13: 978-1-951115-97-5

Edited and designed by Recoil Staff

Printed in the United States of America

10 9 8 7 6 5 4 3 2 1

TABLE OF CONTENTS

In just about any nation other than the USA, the topic of making a firearm quieter is about as uncontroversial as whether or not to play AC/DC at full volume outside a nursing home, or let off fireworks on Sunday in a church parking lot. It's just not acceptable in polite society, and reducing the noise your gun makes is simply the civilized thing to do.

Even in the country in which I was born, with some of the most draconian firearms laws on the planet, it's much easier to obtain a suppressor than it is to own the most mundane gun you could think of. In that case, their overbearing health and safety laws triumph over the myths pushed by Hollywood — hearing damage is very real, very quantifiable, and very permanent. Anyone who has used a rifle on a two-way range is suffering the effects of 180 decibel gunshots, and we currently have an entire generation of servicemembers for whom the first thing they hear in the morning, and the last noise they hear at night is the high-pitched screech of tinnitus.

Suppressors make sense in so many applications, too. If you're teaching a new shooter the basics of marksmanship, it's so much easier (and safer) if you can both communicate without being encumbered by muffs or plugs. If you're hunting, the decision as to whether or not to submit to permanent hearing loss by not wearing ear pro, or possibly miss hearing an animal sneak through the woods should be a no-brainer. And for home defense, should you ever be forced to use a gun to protect your family, then quite apart from the emotional and financial ramifications of that decision, you're going to suffer the aftereffects of touching off a centerfire caliber in a confined space.

At RECOIL, we're unabashed proponents of suppressor ownership and use. Yes, there are bureaucratic hoops to jump through, and a frankly unconstitutional $200 tax to pay, but on balance, the benefits are worth it. For anyone considering buying their first can, but are put off by the waiting period, we always ask, "Will you still want a suppressor in six months' time?" If the answer is yes, then you've just knocked down that obstacle. So come on down, join the NFA party — and if you're looking for a reason to get a second or third suppressor, we're here to help.

Iain Harrison

Editor, RECOIL Magazine

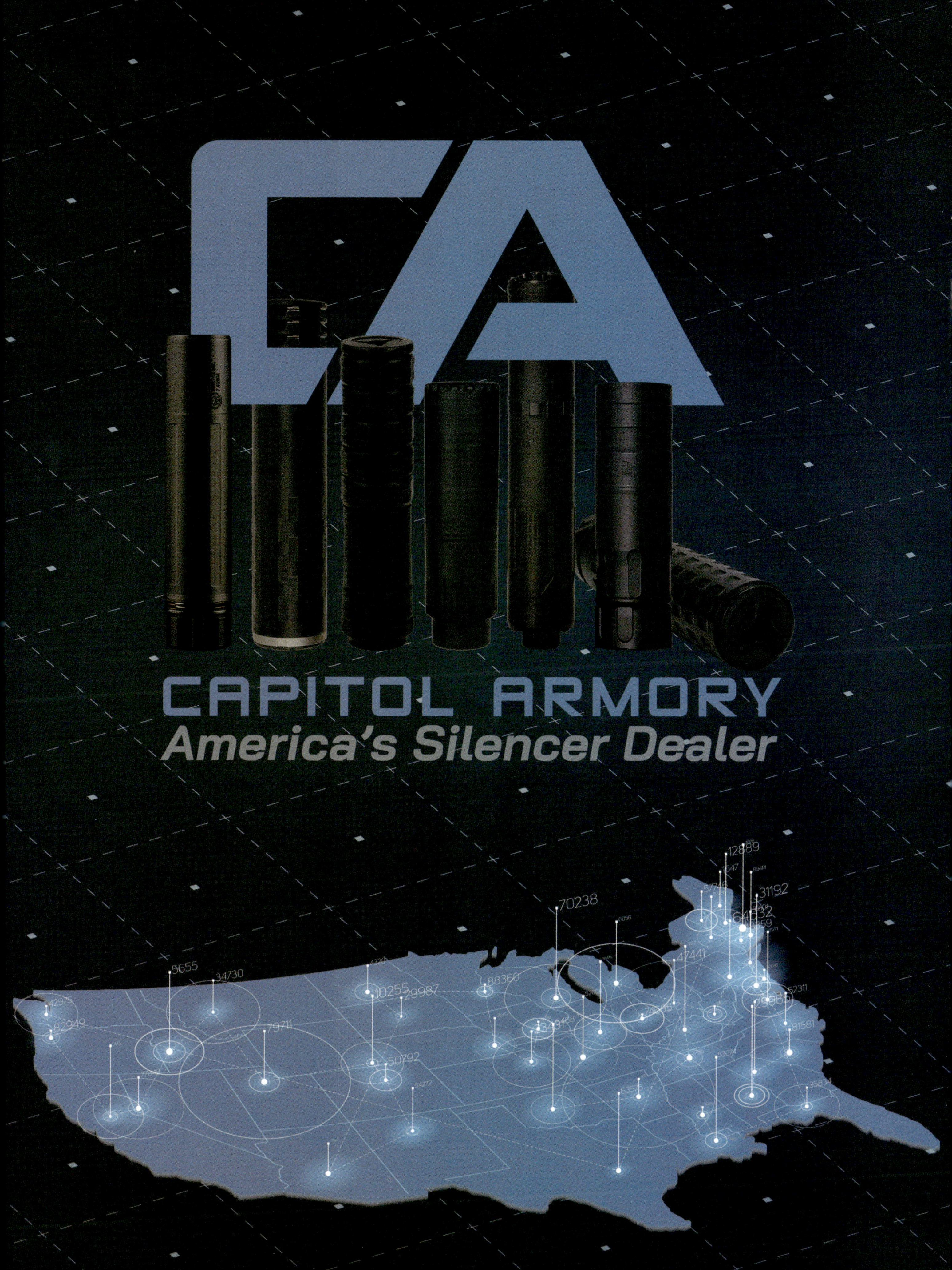

CA
CAPITOL ARMORY
America's Silencer Dealer

CAPITOL ARMORY
America's Silencer Dealer

ABOUT THE COVER

In the world of firearms and shooting sports of any discipline, no product category is as stigmatized and misunderstood as suppressors. Branded largely by Hollywood and misinformed media platforms, suppressors (call them silencers or cans, if you prefer) have been ill-labeled as tools designed specifically for assassins, the military and those "tactical guys."

The irony, of course, is that no other product category can and does improve the shooting experience—for hunters, target shooters and general gun enthusiasts of any skill level—like suppressors.

In 2009, Torrey and Rhiannon Eltiste opened the doors of Capitol Armory with the goal of debunking the myths associated with suppressors, and to make suppressor ownership as simple and efficient as possible. During the next decade, thanks in large part to the efforts of people like the Eltistes, a supportive firearms industry and a massive information campaign, suppressors were successfully removed from the shadows and seen for what they truly offer: unmatched hearing protection, recoil reduction and a new means of enjoying the entire shooting process.

"When Capitol Armory first started, suppressors were largely owned by middle-aged, middle-class men with disposable income," says Randall Durham, Capitol Armory's Director of New Business Development. "But that's not the case anymore. There's so much more information available than there was just a few years ago. And, oddly enough, video games—where a player can not only choose a silencer, but a specific silencer of a specific brand—have opened up the world of suppressors to an entirely new group of shooters."

Early in 2021, Torrey Eltiste reached out to Durham with an idea: to help consumers better understand and obtain suppressors, regardless of where they lived. In short, to become America's silencer dealer. A simple concept? Yes. A huge undertaking? Absolutely.

"I can proudly say that, today, Capitol Armory offers a to-your-door service ... and we will walk with you every step of the way to get there," adds Durham.

"Presently, we hold FFLs in 38 of the 42 states where suppressor ownership is legal, and we're currently working on securing FFLs in those other four," Durham says. "From research through delivery, we take care of everything, and you never have to leave your home."

Capitol Armory employs an extensive staff of suppressor-owning experts ready to field calls or emails from people completely new to suppressors, to those who already hold multiple tax stamps. When a purchase is made, Capitol Armory mails a fingerprinting kit directly to your home, and everything from there forward is completely digital through electronic forms and emails.

When Capitol Armory first opened its doors 14 years ago, the waiting period from purchase to delivery was 90 days or less. That reality is a bit different today.

"The ATF's eForm 4 system offers a lot of hope and promise for our segment of the industry," says Durham. "So much of our world is cloud-based these days, so there's hope that this new system moves things faster. But I'm proud of us as an industry: We've done a good job leveling the playing field about what to expect from the ATF. And when the customers know the truth, it's much easier for everyone to manage their expectations."

"We want everyone to know that we're here to help," adds Durham. "We've built relationships with each and every major manufacturer currently producing suppressors, which means that not only do we have a tremendous number of suppressors on-hand, but we often know what's coming next."

"There's never a middle man when you work with us, and we never do any bait-and-switch sales," he continued. "Our entire team is dedicated to making the non-enthusiasts suppressor enthusiasts. We can literally help with anything."

1
SUPPRESSORS
SOUND
SUPPR
APS 4GC Cal. 9mm
D12A000007
Made in Virginia Beach, VA
GEMTECH
BOISE, ID
D12-54898
G5 5.56mm
Other Patents
Applied For

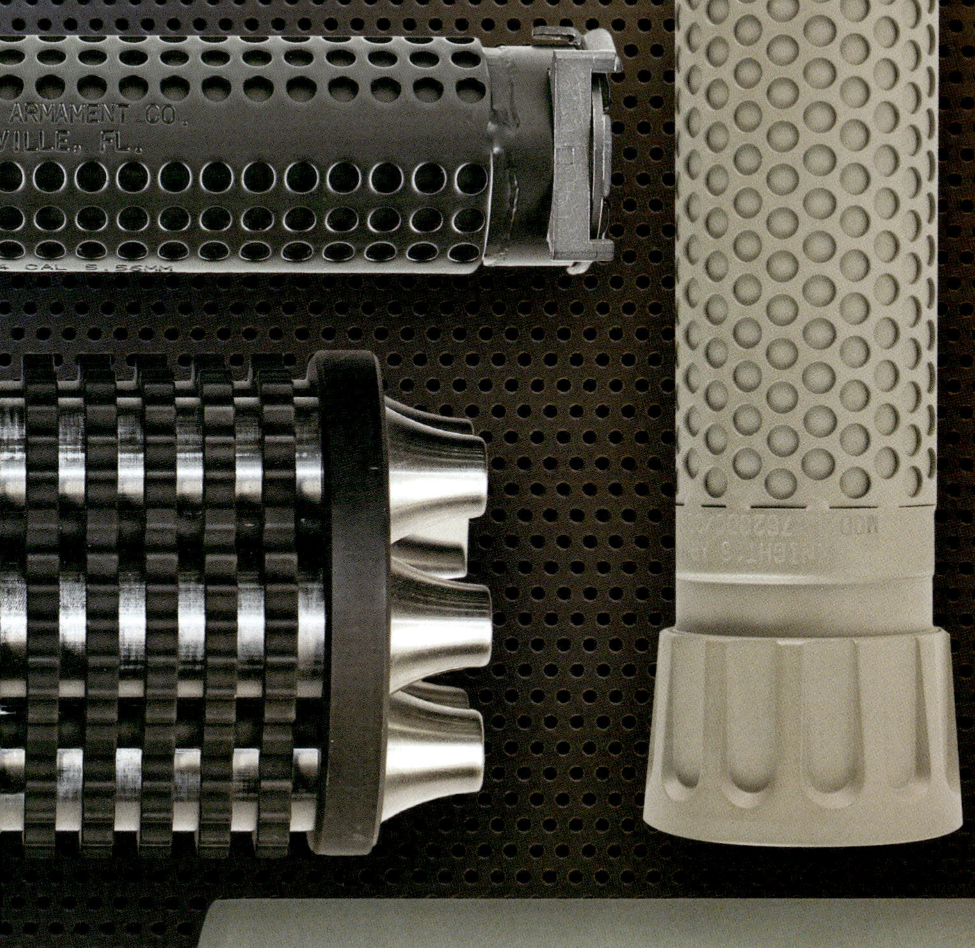

Sound suppressors might be one of the most misunderstood firearms accessories on the market. Everything from their application, functionality, and even legality are questioned and often answered with incorrect information. With the amount of misinformation floating around, it is understandable why it can be a confusing topic. We admit that we have a few questions about suppressors ourselves. That's why we approached the best in the business to help answer some of the most frequently asked questions about sound suppressors. RECOIL brought together a panel of experts from Advanced Armament Corporation, Gemtech, Knight's Armament Company, Silencerco, Spike's Tactical, SureFire, and Yankee Hill Machine Company to help provide definitive answers to the most popular, sound suppressor–related questions.

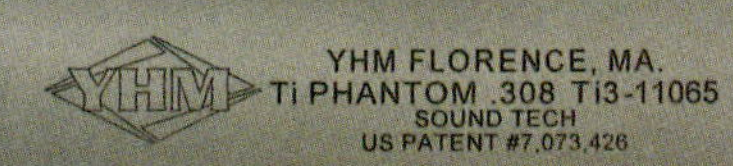

SSORS

Top 20 Questions Answered

By Recoil Staff

GEMTECH
BOISE, ID
D12-52044
TUNDRA-SV 9mm

SUPPRESSOR OR SILENCER?

Let's clarify the terms sound suppressor and silencer. These interchangeable terms are used for devices that reduce the report of or disguise the sound of a gunshot. The Bureau of Alcohol, Tobacco, Firearms and Explosives (ATF) calls these devices silencers. Therefore, the language in the National Firearms Act (NFA) refers to such devices as silencers. Perhaps, this is because in the early 1900s, Hiram Percy Maxim, the inventor of the first commercially available device called them just that, the Maxim Silencer. Because these devices generally don't actually silence a gunshot but suppress its sound, it is agreed among many that the more technically accurate term is sound suppressor. In general, both terms are used interchangeably.

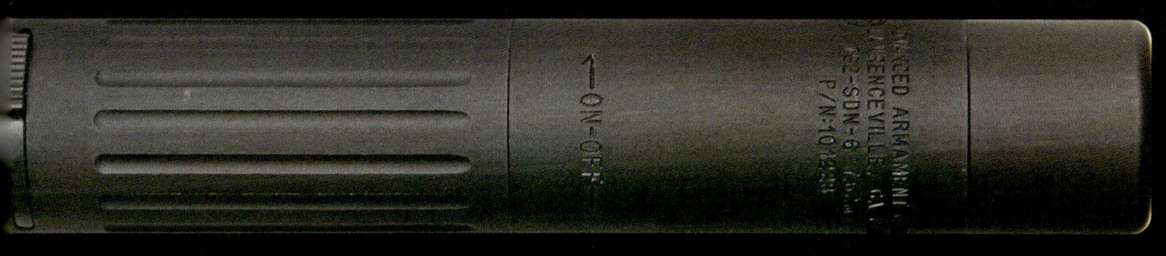

BRAND: SureFire
MODEL: SOCOM 556 Mini
MSRP: $1,299
URL: www.surefire.com
NOTES: Quick Detachable, 5.56mm caliber, for rifle applications

BRAND: SureFire
MODEL: SOCOM 762 Mini
MSRP: $1,399
URL: www.surefire.com
NOTES: Quick Detachable, 7.62mm caliber, for rifle applications

BRAND: SureFire
MODEL: SOCOM 762
MSRP: $1,799
URL: www.surefire.com
NOTES: Quick Detachable, 7.62mm caliber, for rifle applications

TOP 20 QUESTIONS

01 What do sound suppressors do?

Spike's Tactical (ST) explains, "Sound Suppressors help to control the release of pressure coming out of the barrel of the gun, giving it more volume and time to cool as it enters the surrounding lower pressure atmosphere. The sound of the gunshot is created by the high-pressure gas generated by the gunpowder igniting in the round escaping behind the bullet as it leaves the barrel, much like the pop a cork makes leaving a champagne bottle. The suppressor has much more volume for the gas to expand into and cool, so when the gas leaves the suppressor itself the pressure is much lower, therefore creating much less sound." Gemtech (GT) adds, "Sound suppressors, at their simplest, are mufflers designed to take this high-pressure, high-speed event, and slow it down from a fast, loud pop into a slower, muffled *thump*."

02 What are sound suppressors used for?

SureFire (SF) points out that, "Using a firearm without hearing protection can result in permanent hearing loss after just one shot. The main purpose of a suppressor is to protect a shooter's hearing and the hearing of those around them." Silencerco (SC) elaborates, "The suppressor will also typically reduce felt recoil, which is a great way to start new shooters to our sport without all the intimidating and bad habit–inducing negatives of muzzle report, blast, and recoil. The suppressor makes shooting fun for new people, instead of a scary event. It is great for training kids and adults, who will get much more enjoyment and better shooting performance out of using them, as well." From the training aspect, GT mentions that "increased ability for shooters to hear during training and instructors not having to scream commands over gunfire are obvious benefits." SC speaks about their use in military and law enforcement. "[T]hey offer the same advantages as they do to civilians with the added bonus of greatly reducing or eliminating muzzle flash and dust signature. This makes our soldiers and police officers less of a target and increases their survivability."

03 Who would want to use a sound suppressor and why?

GT says, "There's never a good reason to go deaf." We have to agree with that! SF continues, "Suppressors can benefit everyone, including hunters and target shooters. Unlike in the movies where a suppressor eliminates virtually all the noise from a gunshot, most rifle ammunition will still produce a ballistic crack, but it will be at a safe level. Hearing damage is very common in the military, and use of suppressors helps mitigate this. For all of these reasons, suppressors make sense in every shooting situation."

BRAND: Spikes Tactical
MODEL: Suppressor- LRS-1
MSRP: $999
URL: www.spikestactical.com
NOTES: Direct thread, 5.56mm caliber, for rifle applications

BRAND: Spikes Tactical
MODEL: Suppressor- MRS-1
MSRP: $699
URL: www.spikestactical.com
NOTES: Direct thread, 5.56mm caliber, for rifle applications

04 What are key pieces to know about suppressor purchase and ownership?

Advanced Armament Corporation (AAC) answers, "Silencers are legal for civilian ownership in 39 states (and counting). Contrary to popular belief, silencers are and always have been legal to own under federal law. Getting one takes just two simple forms. It's as easy as locating a dealer in your state that sells silencers, filling out and mailing the appropriate forms to the BATFE, waiting patiently, and then picking up your new can and start enjoying the countless benefits of quieter shooting. The paperwork associated with ownership never needs to be renewed. It is a onetime, lifetime registration per silencer." GT adds some details, "Getting a suppressor as an individual is simply a waiting game, and $200 [for the NFA tax stamp]. It's not as complex as some make it sound, it's not giving up any rights... it's a title change, just like a car title changing hands, only it's done through an antiquated system, dating back to 1934 that predates computers and instant background checks. But, if you have a clean background that wouldn't prevent you from owning a pistol, then you're good to go. Just know you won't be taking it home from the gun shop that day — a good dose of patience and the price of a nice dinner out will get your transfer done, and then the suppressor is yours; no annual fees, no inspections, no giving up of any rights. The suppressor isn't married to any one gun; you can move it from the oldest rifle in the gun safe to whatever the newest firearm to come out might be."

05 What are the different types of suppressors for pistols and rifles?

SC informs us, "There are suppressors for virtually any caliber and platform with the exception of revolvers. The typical categories for suppressors are: .22 Rimfire (.22LR/. 22WMR), Centerfire Handgun (9mm, .40 S&W, .45 ACP), Centerfire Rifle (5.56/.223 and 7.62/.308) and large bore (.338 and .50). There are many different calibers available, but these are the calibers and categories most often encountered in the market place."

06 What are Wet and Dry suppressors?

Wet. Dry. What? AAC describes the difference to us, "Most modern silencers are designed to be fired in a dry condition. This means that they require no performance enhancers, such as water, grease, or oil, in order to achieve a reasonable level of sound reduction. However, most centerfire pistol designs will exhibit less muzzle and ejection port flash, and a reduction in overall sound signature if the interior of the silencer is treated with a small amount of ablative media, such as plain water. Shooting a handgun silencer wet is as simple as introducing about 5 cc of simple water or other ablative materials, such as wire pulling gel, ultrasonic gel, or other coolants into the rear of the silencer. The sacrificial media typically remains effective for one to two magazines of use before it must be replenished." Rifle suppressors, however, are not recommended to be run wet. GT states, "Putting coolant into a rifle suppressor isn't a good idea, as there's already so much pressure from the expanding gunshot gases that you could damage the suppressor — if all the chambers of the suppressor where the gas would normally expand into are already taken up with non-compressible water, the tube could bulge."

BRAND: Advanced Armament Corporation
MODEL: MINI4
MSRP: $895
URL: www.advanced-armament.com
NOTES: Quick Detachable, 5.56mm caliber, for rifle applications

BRAND: Advanced Armament Corporation
MODEL: 762-SDN-6
MSRP: $1,050
URL: www.advanced-armament.com
NOTES: Quick Detachable, 7.62mm caliber, for rifle applications

07 What are the basic parts of a suppressor and what does each part do?

Let's see what a sound suppressor's made of. AAC breaks it down, "All silencers feature an expansion or blast chamber that the bullet must travel through before making its way through each subsequent baffle and eventually exiting the silencer. The expansion chamber is generally the chamber with the largest volume within the silencer to allow for the initial introduction of the hot expanding gases propelling the bullet. These hot expanding gases eventually make their way through the baffle stack, consisting of a designated number of baffles and spacers, which are essentially smaller chambers designed to disrupt the natural path the gas would take. By the time the gases exit the silencer, they have slowed considerably and produce a quieter sound signature."

This is what is inside of an AAC Ti-RANT pistol silencer. The Ti-RANT series of silencers are available in 9mm and .45 calibers.

SOUND SUPPRESSOR ADVANTAGES & DISADVANTAGES

Our panel of sound suppressor experts universally agrees that suppressors are a good thing. Shocking news, isn't it? Aside from the obvious benefits of running a sound suppressor like protection from irreversible hearing loss, we asked them to name a few more specific reasons. This is what they answered.

ADVANTAGES

"Silencers reduce recoil and muzzle flip, allowing for more accurate and faster follow-up shots." – **ADVANCED ARMAMENT CORPORATION**

"There's never a good reason to go deaf, especially from fun activities like shooting, where firearms-mounted hearing protection can open up where and when you can shoot without disturbing others. There's nothing like spending a weekend putting hundreds of rounds downrange and not feeling like you've been trapped in hot, sweaty earmuffs all day." – **GEMTECH**

"[Suppressors] allow first-time shooters the chance to enjoy shooting without the potential of being scared off by the noise." – **KNIGHT'S ARMAMENT COMPANY**

"At its most basic, reduced muzzle report, reduced recoil, reduced flash, a decrease in shooter fatigue, and an increase in shooter enjoyment and performance." - **SILENCERCO**

"Reduced muzzle flash of the host firearm." – **SPIKE'S TACTICAL**

"They reduce the felt recoil of a firearm by as much as 30 percent." – **YANKEE HILL MACHINE COMPANY**

DISADVANTAGES

Of course, nothing super awesome comes without costs. In the case of sound suppressors, the monetary cost besides the suppressor itself is a $200 tax stamp. Then, there are the length and weight that a suppressor brings to the end of your firearm. They also get your gun dirty much quicker than firing it without one, and they can cause a shift in your point of aim, point of impact. But, as YHM indicates, there is one disadvantage that we don't mind dealing with, which is "an increase in smile-related muscle fatigue may be noticed." Sign us up!

BRAND: DeGroat Tactical Armaments, LLC.
MODEL: XM134-SD
MSRP: contact manufacturer
URL: www.armamentsales.com
NOTES: 7.62mm caliber suppressor for Minigun applications. Yes, that's right, Minigun.

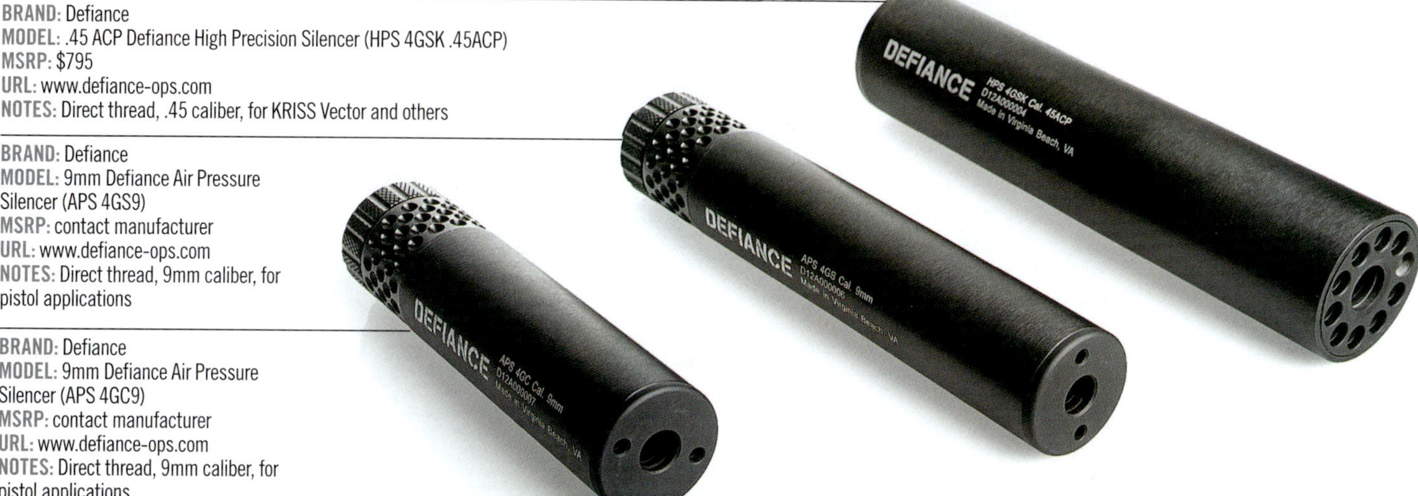

BRAND: Defiance
MODEL: .45 ACP Defiance High Precision Silencer (HPS 4GSK .45ACP)
MSRP: $795
URL: www.defiance-ops.com
NOTES: Direct thread, .45 caliber, for KRISS Vector and others

BRAND: Defiance
MODEL: 9mm Defiance Air Pressure Silencer (APS 4GS9)
MSRP: contact manufacturer
URL: www.defiance-ops.com
NOTES: Direct thread, 9mm caliber, for pistol applications

BRAND: Defiance
MODEL: 9mm Defiance Air Pressure Silencer (APS 4GC9)
MSRP: contact manufacturer
URL: www.defiance-ops.com
NOTES: Direct thread, 9mm caliber, for pistol applications

08 Why do I see so many different kinds of mounting options? Are they maker specific?

Knight's Armament Company (KAC) describes various mounting options, "Without addressing legal issues, the mounting devices seen on rifle suppressors are proprietary mounts that typically allow a better mounting platform than standard threads. Every suppressor manufacturer has their own muzzle device for mounting their suppressor(s). These devices can be flash hiders, muzzle brakes, or simply a slight modification to the end of the barrel to allow for their suppressor to be mounted. Some makers design systems to be caliber specific, while others will make their suppressors and muzzle devices capable of being mounted on multiple weapons or calibered weapons."

09 Direct thread versus Quick-Detach (QD) mounting, is there a difference?

SC points out that there is a difference. "A direct thread suppressor screws on directly to a host weapon's barrel threads. A QD version usually has a flash hider or other type of mount that stays attached to the host firearm at all times and the suppressor then mounts to the attachment device or flash hider. It really boils down to personal preference as to which is better or more suitable for any given application." GT mentions, "The lightest, shortest, simplest way to attach a silencer to a rifle is by threading it directly to the barrel, and we're seeing more and more professional users starting to see that quick disconnect is less important, instead valuing lighter, shorter, less expensive direct-thread suppressors." Direct thread is a less expensive way to go, as well.

10 How do flash hiders affect a suppressor? How about muzzle brakes?

"Flash hiders have no negative effect on the silencer, but muzzle brakes can actually extend the serviceable life of your can. The blast chambers on our muzzle brakes act as a sacrificial blast baffle and can take the brunt of the weapon's muzzle blast. There is no sound difference between flash hiders, muzzle brakes, or flash-suppressing muzzle brakes with the silencer mounted. However, muzzle brakes do tend to produce more noise to the shooter than flash hiders without a silencer attached," explains AAC. KAC offers a different evaluation, "Flash hiders and muzzle brakes react differently within a suppressor, but this is another situation where it is more dependent on the manufacturer and the technology they utilize in their systems. Typically, a flash hider will perform better in a suppressor as compared to a muzzle brake for first round flash. A flash hider is designed to do exactly as its name indicates, reduce flash. A muzzle brake is designed, primarily, to redirect the exiting gases in a manner to allow the shooter to experience less recoil or muzzle movement. This redirection may cause more flash to come out the front of the suppressor on the first round compared to the flash hider. But, as previously stated, this is more dependent on the manufacturer and their technology."

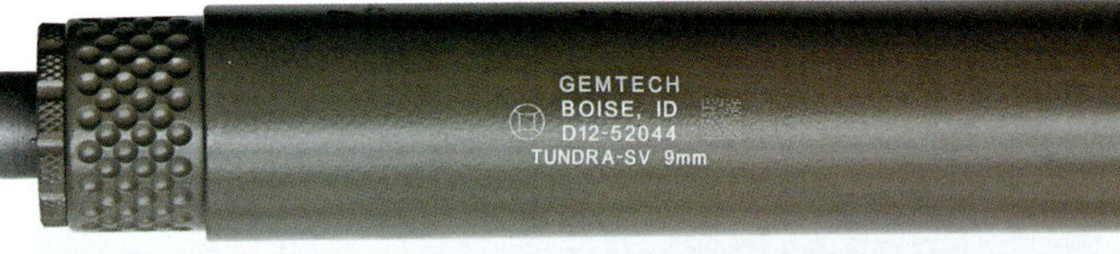

BRAND: Gemtech
MODEL: Tundra-SV
MSRP: $750
URL: www.gem-tech.com
NOTES: Direct thread, 9mm caliber, suited for compact pistols

BRAND: Gemtech
MODEL: G5
MSRP: $875
URL: www.gem-tech.com
NOTES: Quick Detachable, 5.56mm caliber, for rifle applications

BRAND: Knight's Armament
MODEL: M4/M16 QDSS-NT4
MSRP: contact dealer
URL: www.knightarmco.com
NOTES: Quick Detachable, 5.56mm caliber, for rifle applications

BRAND: Knight's Armament
MODEL: QDC (Quick Detach Coupling)
MSRP: contact dealer
URL: www.knightarmco.com
NOTES: Quick Detachable, various calibers and sizes for rifle applications

11 How does caliber affect suppressor choice?

ST informed us that, "Typically, the larger the caliber the larger the suppressor needs to be for that caliber to be effective. The suppressor for a larger caliber will also need to have a sturdier construction to handle the increased pressure from a larger caliber. A centerfire rifle caliber suppressor will be larger than a centerfire pistol caliber suppressor, and both will be larger than a rimfire caliber suppressor, typically." ACC also explains, "Certain calibers are much easier to effectively suppress than others. Any bullet traveling fast enough to break the sound barrier will cause a sonic crack and will not be Hollywood quiet. The speed of sound is roughly about 1,050 feet per second at sea level. While the silencer will still slow and cool the gases prior to releasing them into the atmosphere and the overall sound signature of the combustion gases will still be reduced, the sonic crack of the round traveling through the air will produce a sound in the high 130dB range. Some of the most commonly available rounds that are just as quiet as the movies are .22LR, 9mm (147 gr. subsonic), and the new 300 AAC Blackout (220 gr. subsonic). If you are seeking the greatest overall sound reduction, stick with one of the aforementioned calibers and try not to walk away smiling."

12 Can I run a larger caliber suppressor on a smaller caliber gun? What are the positives and negatives to a setup like that?

AAC starts us off by saying, "As a general rule of thumb, you can use a larger caliber silencer on a smaller caliber firearm, provided you follow some simple guidelines. There are three categories of silencers: rimfire, centerfire handgun, and centerfire rifle. As a general rule, stay within the same category when using your silencer on an alternative caliber. You can use a silencer on an alternate caliber, as long as the projectile diameter and pressure are equal to or less than the caliber the silencer was designed for." SF adds, "One thing to keep in mind is a larger bore will allow the gases to evacuate faster, but the typical longer length of the larger bore suppressors can counteract that and allow the gases to cool before existing at high velocity." KAC warns us about performance issues, "Usually, running a larger caliber suppressor on a smaller caliber firearm won't cause any issues with the suppressor. This is something commonly seen done on handgun suppressors. The larger suppressor will typically be a few decibels (dBs) louder, when shot with a smaller caliber round than its intended design. When it comes to rifle caliber suppressors, there is more potential for an issue. For instance, a person may want to shoot a .22LR through a 5.56 suppressor and, while the rounds are nearly the same diameter, they do completely different things within a suppressor. Also note when shooting a small caliber in a larger suppressor, the performance of the suppressor may be diminished, aside from the increase in dBs; how much the performance drops depends on the manufacturer and technology employed."

BRAND: Ops Inc.
MODEL: 3rd Model
MSRP: contact manufacturer
URL: www.opsinc.us
NOTES: .30 caliber, for use on rifle applications. Shown here on an M240B. Yes!

13 We sometimes see different length suppressors for the same application. Does length affect sound suppressor effectiveness?

SC says, "Length, or to be more specific, internal volume does play a role. You can compensate for some of that through baffle stack design. The user needs to determine the right balance in size, weight, and performance. Sometimes, the biggest suppressors are not always the quietest or best performing in other areas. Design is a huge factor, and size and weight are major decision criteria for most users." SF confirms this, stating, "The baffle stack design and gas flow dynamics are the determining factors of a suppressor's effectiveness, but length of a suppressor does help if the first two criteria are equal. For example, an 8-inch suppressor with a superior design will be more effective than a 12-inch suppressor with poor gas flow dynamics."

14 How does the heat generated from firing a gun affect the suppressor?

"The heat that is generated degrades the internals of the suppressors." ST continues, "Modern suppressors have internal components made of materials designed to withstand the high temperatures of sustained gunfire. As the temperature increases inside of the suppressor the suppression efficiency degrades slightly. Yankee Hill Machine Company (YHM) goes on to say, "The size and type of round being fired, barrel length, and suppressor material all play a role in just how hot the suppressor gets and how quickly it gets hot."

15 What kind of cycling issues might I have with using a suppressor?

"Rimfire silencers and locking breach systems, like bolt or break open rifles are typically issue-free when using a silencer," explains AAC. "Handguns with 'Browning tilting barrel' type systems need a Neilson-type device for 100-percent reliability. Many modern piston-type rifle designs incorporate a Silencer setting, due to increased backpressure. Some consumers prefer an adjustable gas system, such as the Noveske Rifleworks Switch Block, to tune their rifle for less backpressure and softer function, but it's not a function issue, more a shooter comfort issue. There can be ammunition cycling issues with specialty ammunition, like some subsonic rifle ammunition that is not designed to cycle. Specialty subsonic cycling rounds in 5.56 mm tend to be very expensive and reliability can be spotty. These issues were of primary importance in the design of the 300 AAC Blackout round. The Remington 220-grain Subsonic Sierra OTM round was designed to provide reliable semi- or full-auto performance, while costing a fraction of the price that you pay for less reliable, specialty cycling subsonic."

16 What are boosters and Nielsen Devices?

YHM clarifies this for us, "A booster or Nielsen Device is a mechanism that basically allows for the weight of the suppressor to be relieved from the host firearm after firing and during cycling. The system is a piston on a spring that allows the suppressor to float while the pistol unlocks and cycles. Without this, the additional weight of the suppressor would not allow the actions of most pistols to open and cycle properly."

17 Does ammunition choice affect sound output? What are subsonic rounds?

ST confirms, "Yes, ammunition has an effect on sound output. Typically, the hotter a round is loaded the more sound pressure waves are generated. And with supersonic [1,050 fps or so depending on altitude] rounds the crack of the round breaking the sound barrier always has an effect on sound output. Subsonic rounds are loaded in a manner that the bullet doesn't break the sound barrier, not having the crack of breaking the sound barrier helps reduce the sound output." SF backs this up with a little math, "Any round that exits the bore above the speed of sound, which is 1,118 fps when measured at sea level, 59 degrees F, 29.53 in hg, and 78 percent humidity, will break the sound barrier. The resulting noise, essentially a small-scale sonic boom, is commonly referred to as the ballistic crack. Subsonic ammunition exits the bore under that velocity threshold, and hence does not produce a ballistic crack."

18 Does ammunition choice affect reliability of using my gun with a suppressor? If it does, how so?

SF answers, "Subsonic ammunition can adversely affect weapon functioning, since it does not have the gas pressure required to cycle the bolt. Hot-loaded ammunition can cause the suppressed weapon to malfunction, due to an excessively high gas flow that cycles the bolt faster than the magazine can feed the rounds. Both of these cases are rare and can usually be fixed by using good-quality standard-pressure ammo." SC sums it up, "Just like host firearms, some ammo may work better in your particular gun and suppressor combination. You'll just have to try several and see which works best for you." ST reminds us that, "Frangible ammo should not be used with a suppressor. Use of good-quality ammunition should not have any adverse affects."

BRAND: Yankee Hill Machine
MODEL: .22 Mite Sound Suppression System
MSRP: $299
URL: www.yhm.net
NOTES: Direct thread, .22LR caliber pistol applications

BRAND: Yankee Hill Machine
MODEL: Titanium Q.D. Phantom 7.62
MSRP: $1,100
URL: www.yhm.net
NOTES: Quick Detachable, .30 caliber rifle applications

19 Is using a suppressor harsh on the gun's system?

YHM explains, "Some users may notice increased wear on certain parts of their semi-automatic firearm, when using a suppressor. The suppressor causes a pressure increase in the cycling mechanism that generally causes the firearm to cycle more quickly." That being said, SC points out that "the only real disadvantage to using a sound suppressor is it will make your host firearm dirtier more quickly. It typically won't result in any noticeable premature wear on your gun." SF adds, "Suppressor use requires a little bit more frequent cleaning on gas-operated guns. Suppressor use on a gas-operated gun also tends to dry the bolt out a little faster, which means you need to oil it more. Other than that, a suppressor does not have any adverse effect."

20 How much sound can I expect to reduce from my gun?

"Awesome" is all we said when we heard GT say, "If you go into suppressors thinking it's going to sound like a James Bond movie, you might be disappointed. If you go into it understanding a quality suppressor is going to be better than plugs or muffs, changing a sharp, loud crack into a muffled thump that won't ring your ears, you'll be pleasantly surprised. Also, .22LR silencers are definitely the gateway drug of suppressors — they sound as close to the movies as the industry can give you and are really what people expect out of a silencer, but any caliber firearm benefits from being made quieter for the shooter and bystanders."

ADVANCED ARMAMENT CORPORATION
www.advanced-armament.com
www.aaccanu.com

GEMTECH
www.gem-tech.com

KNIGHT'S ARMAMENT COMPANY
www.knightarmco.com

SILENCERCO
www.silencerco.com
www.silencersarelegal.com

SPIKE'S TACTICAL
www.spikestactical.com

SUREFIRE
www.surefire.com

YANKEE HILL MACHINE COMPANY
www.yhm.net

LIBERTY SUPPRESSORS' MYSTIC X

The Swiss Army Knife of Cans

By Iain Harrison

Photos by Kenda Lenseigne

One of the most daunting tasks facing anyone who wants to enter the world of NFA goodies is picking that all-important first suppressor. While there's an excellent case to be made that one should bust one's cherry on a decent .22 rimfire can, the Mystic X is one of the most versatile silencers currently available — just about any caliber you can run through an AR-15 can be suppressed by this unit. Although it's quite a step up in terms of purchase price, the ability to be employed on centerfire rifles, handguns — and yes, .22 LR guns — makes it worthy of your consideration.

CONSTRUCTION

Employing a stainless-steel monocore baffle stack, the Mystic X can be fairly easily disassembled for cleaning. While this is no real advantage in a dedicated rifle can, .22 rimfire (and to a lesser extent, 9mm) is notoriously filthy and will bung up a suppressor in no time if left to its own devices. In order to prevent the tube from filling up with caked-on lead cack, it's necessary to

MAKE:
Liberty Suppressors

MODEL:
Mystic X

CALIBER:
9mm and smaller

LENGTH:
8 inches

DIAMETER:
1.375 inches

WEIGHT:
10.5 ounces

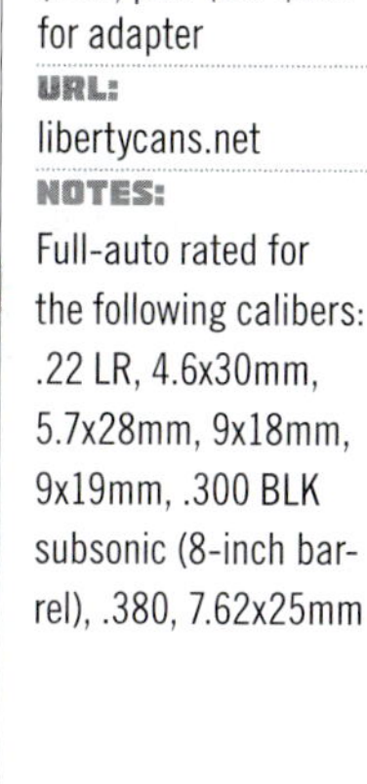

MSRP:
$799, plus $95-$155 for adapter

URL:
libertycans.net

NOTES:
Full-auto rated for the following calibers: .22 LR, 4.6x30mm, 5.7x28mm, 9x18mm, 9x19mm, .300 BLK subsonic (8-inch barrel), .380, 7.62x25mm

clean the baffles periodically — next to impossible in welded-core designs. To clean the Mystic X, the user simply unscrews the caliber adaptor and lock ring, then pushes the core out of the titanium outer tube. Once free, it can then be scrubbed, blasted, or dipped in an ultrasonic tank.

A variety of adapters are available in order to convert the suppressor to attach to numerous firearms in your safe. Ours arrived with ⅝-24 and ½-28 fixed barrel adapters, as well as a pistol adapter equipped with a Nielsen device to permit functioning on locked breech, recoil-operated handguns. We lost no time in screwing it onto as many guns as we could in order to test the manufacturer's versatility claims, although we didn't manage to try it out on every one of the 60 calibers it's rated for. Which is an excellent reason to buy more guns.

GOING HOT

Weighing in at just shy of 11 ounces without an adapter, the lack of thermal mass is quickly noticed after firing a few rounds of full-power rifle ammo — this can heats up quickly. It's a good idea to check any direct-threaded suppressor after five or six rounds to make sure it hasn't started to shoot loose. When we did, the Mystic X was pretty toasty. Unlike a dedicated 5.56mm suppressor such as AAC's benchmark M4-2000, this unit doesn't have an inconel blast baffle to take the hit from emerging gasses. So the Mystic X is limited to longer barrels and a slower rate of fire (except those listed in the specs) — screw it onto your 10-inch SBR and the unburned powder emerging at 12,000 psi will erode it in short order, but follow the manufacturer's instructions and it'll last a long time.

Compared to some other cans, it's pretty long, but the upside to this is greater internal volume, hence better noise dampening. On our Brethren Armament MP5 clone, it's stupid quiet with 147-grain subsonic ammo. In .22 LR, despite the big gap between the projectile and baffle stack, it's at least as quiet as a dedicated .22 muffler, even if it does look a little out of place on a Ruger 10/22.

Our only niggles were that after a while, disassembly becomes something of a chore due to carbon buildup. We had to heat up the adapter with a blowtorch in order to break it free, and the lock ring needed more of the same, plus the gentle ministrations of a set of Vise-Grips. Apart from that, it lives up to the manufacturer's claim and is one of the most versatile suppressors currently on the market. **R**

YES, YOU CAN

RYDER 9 TI: SUREFIRE'S FIRST PISTOL SUPPRESSOR

By Iain Harrison
Photos by Kenda Lenseigne

We shot SureFire's SOCOM 2 suppressor on the Accuracy International rifle featured on page 72 and were impressed by its performance. So, when their rep asked if we'd be interested in getting some time with their newest model, there was very little arm twisting involved.

Bottom line up front: it's a modified K baffle design, using a titanium tube and end caps for reduced weight. The baffles themselves are stainless steel, which should mean they last just about forever, and in the event of erosion or baffle strikes can be individually replaced by the manufacturer. SureFire claims that flats machined on the outer tube reduce weight and allow for a better sight picture. The first part of that sentence is demonstrably true (although whether the savings are significant or noticeable is up for debate). We were unable to verify the second in the limited amount of time we spent with it. The Salient Arms Glock 17 we used as a host was equipped with both a Trijicon RMR and suppressor-height sights, so lining up the irons was a non-issue.

Throughout our (admittedly short) evaluation, the Nielsen device fitted to the Ryder 9 Ti's end cap worked well — recoil was not intrusive and the gun functioned without a hitch, indicating that the piston was well-tuned to the job at hand. While our 115-grain test ammo was less than ideal, the can did a commendable job of reducing muzzle report and blast, allowing us to shoot without hearing protection. Available in both ½-28 and 13.5x1mm left-hand thread pitches for U.S. and European pistols, SureFire also offers Glock barrels by ZEV Technologies with the same thread options. This means that if you own, say, a SIG SAUER P226 with a factory threaded barrel, you can get the same setup on your Austrian pistol and not need to swap out the mount.

MAKE:
SureFire

MODEL:
Ryder 9 Ti

CALIBER:
9mm

OVERALL LENGTH:
7.6 inches

DIAMETER:
1.25 inches

WEIGHT:
9.5 ounces

MSRP:
$799

URL:
www.surefire.com

Finished in high-temperature Cerakote available in three different colors (black, gray, and dark earth), the Ryder 9 Ti is a good-looking suppressor, and its $800 price point is in line with or a little below comparable offerings from other major muffler companies. **R**

MR. SANDMAN

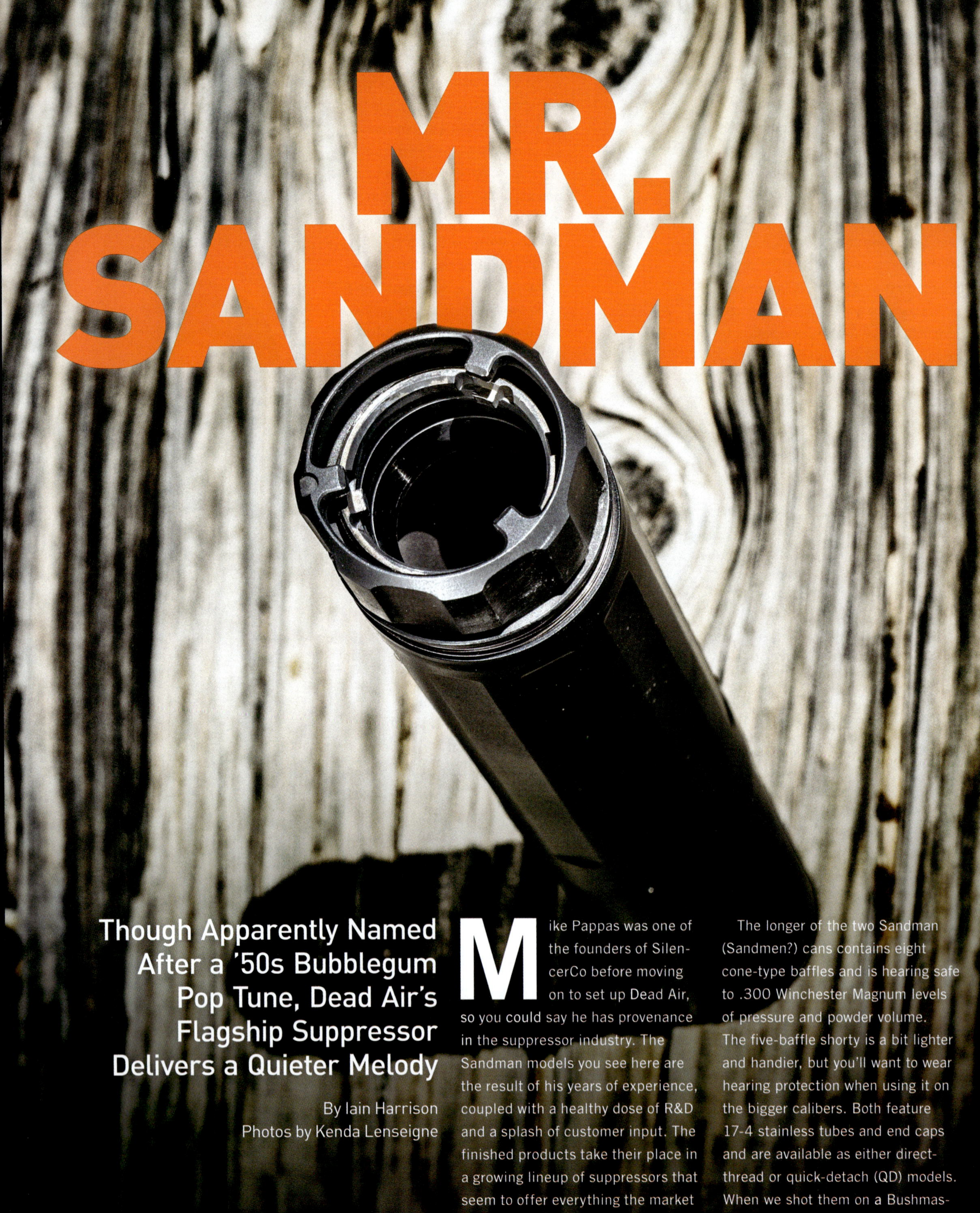

Though Apparently Named After a '50s Bubblegum Pop Tune, Dead Air's Flagship Suppressor Delivers a Quieter Melody

By Iain Harrison
Photos by Kenda Lenseigne

Mike Pappas was one of the founders of SilencerCo before moving on to set up Dead Air, so you could say he has provenance in the suppressor industry. The Sandman models you see here are the result of his years of experience, coupled with a healthy dose of R&D and a splash of customer input. The finished products take their place in a growing lineup of suppressors that seem to offer everything the market has demanded, and then some.

The longer of the two Sandman (Sandmen?) cans contains eight cone-type baffles and is hearing safe to .300 Winchester Magnum levels of pressure and powder volume. The five-baffle shorty is a bit lighter and handier, but you'll want to wear hearing protection when using it on the bigger calibers. Both feature 17-4 stainless tubes and end caps and are available as either direct-thread or quick-detach (QD) models. When we shot them on a Bushmaster ACR, we were impressed by the

MAKE:
Dead Air
MODEL:
Sandman
CALIBER:
Up to .300 Winchester Magnum
OVERALL LENGTH:
6.8 inches (S Model)
8.9 inches (L Model)
WEIGHT:
17.3 ounces (S Model)
21.8 ounces (L Model)
MSRP:
$1,050 (S Model)
$1,200 (L Model)
URL:
www.deadairsilencers.com/products.html

low tonal quality of the muzzle blast, which was among the most pleasant we've encountered.

Instead of utilizing an Inconel blast baffle and then less expensive, easier-to-machine stainless components further down the baffle stack, Dead Air uses Stellite, an extremely erosion-resistant and costly alloy that is typically cast and then ground, rather than machined, due to its toughness.

When used in conjunction with its dedicated muzzle brake, which absorbs the initial jet of superheated gas, there's very little chance that a Sandman will die by erosion. "You couldn't afford the ammo to wear it out, and if you did, you'd have better things to do with your time," Pappas said. Challenge accepted — time to build that 7.5-inch barreled NEMO .300 Win Mag we've always wanted.

"What about baffle strikes?" we hear you cry. Well, the most likely part of the suppressor to be hit by an errant bullet is the end cap, as yaw and lateral errors are magnified by distance from the muzzle. In this case, the end cap is user-replaceable, which is a bonus for those of us who drop expensive stuff on a regular basis.

A word on the Dead Air brake, an example of which ships with every QD suppressor. Apart from acting as the interface between barrel and can, it does a pretty decent job of recoil reduction on its own, without being completely obnoxious. When shot side by side with the Lancer Nitrous compensator on our 14.5-inch barreled three-gun carbine, its reduced concussion was readily apparent. By tapping off a portion of the gasses entering its first expansion chamber and directing them forward, the sideways blast exiting the next three chambers is disrupted, reducing impact on the shooter.

To utilize the QD feature, the can is aligned to the 12 o'clock position and slipped onto the brake — there's only one way it can go on, so point of aim shift is both minimized and, just as importantly, consistent. The shooter then twists the suppressor body to lock the three primary and secondary lugs onto the brake, whereupon the locking ring stops and the tube continues ratcheting for another half turn or so. At this point, the beveled mating surfaces on both can and brake meet to form a self-aligning, gas-tight seal, which keeps carbon away from anything important and allows the user to easily dismount the can, even after many hundreds of rounds. Sounds peachy, right?

We thought so, too. Then we launched a can downrange from the muzzle of a .300 Win Mag. Looks of consternation went around the group, as up till now everything in the field evaluation was going great; from mag dumps on a .223 to ringing steel at 60 yards with a suppressed .22 pistol. Looking at the suppressor, it became apparent that the QD mount had sheared its locking lugs and everyone stroked their chins and offered opinions as to what might have caused the problem. "Not sure what's wrong with it," offered Pappas. "But we're going to test the shit out of it until we find out."

True to his word, he called a week later after discovering that the batch of locking rings arriving the day before our test were compromised by faulty heat treatment. Instead of the RC 40 value originally spec'd, these tested at a butter-soft RC 20 which, while good enough for .223 and .308, failed to withstand the mighty Win Mag. One plane ticket and a case full of ammo later, we were satisfied the problem had been rectified and can now unreservedly recommend the Sandman on calibers up to its rated limit. **R**

SIG SRD 338TI QD

More Acronyms Than You'll Ever Want — for Possibly the Last Rifle Can You'll Ever Need

By Iain Harrison
Photos by Straight 8

Here at RECOIL, we're all about making your dollar go a little further, even if it's only to turn around and spend it on high-end, Gucci kit. Sure, there are times when you need that one, dedicated tool for a very specific job. But if we can get our gear to serve in multiple roles, then that frees up the budget for other important things in life. Like whiskey.

While you may never actually buy a .338 Lapua Magnum rifle, you may nonetheless be interested in SIG SAUER's suppressor tailored for that round — and here's why. Tipping the scales at a very reasonable 19 ounces, it's engineered to take the kind of hammering that a 250-grain bullet at 3,000 feet per second can dish out. As a result, pipsqueak calibers like .300 Win Mag, 8mm Remington Magnum, and .30-06 are a walk in the park. It laughs in the face of .308, and as for 5.56mm — what's that?

So it's lighter than an AAC SDN-6, handles way more calibers, and has enough volume to muffle even the noisiest blasts...you've gotta pay for it with a combination of your left nut, a kidney, and your first-born child, right? Nope, you get five bucks in change from a grand. Although not an inconsiderable amount of scratch, it does mean you cover just about any rifle caliber that's worth shooting. Weight savings come from the material used in its construction, Grade 5 titanium. As important as what it's made from is how it's made — typical cans utilize a welded core that is then slipped inside an outer tube, but in this case, the core *is* the tube and no extra material is required.

To enable your precious purchase to be passed around like a stripper at a frat party (cue outraged social justice warriors in 3...2...1), SIG engineers developed a clever muzzle device that uses a Morse taper to ensure the can stays in place, no matter what host you mate it to. As any machinist will tell you, the Morse taper can be found on lathe tailstocks, end mill collets, and other stuff that grips things very tightly. When torqued, it results in a friction fit between the male and female parts, requiring 20-percent more force to release than to tighten. As a result, once you crank the suppressor onto the muzzle device, you'd better have eaten your Wheaties that morning if you want to unscrew it. The downside to Morse tapers is that they are susceptible to nicks, scratches, and dirt. So don't use your brake as a post hole digger, and if you do, make sure you clean it off before attaching your can.

MAKE:
SIG SAUER Silencers
MODEL:
SRD 338Ti QD
CALIBER:
Multi
OVERALL LENGTH:
10.5 inches
DIAMETER:
1.75 inches
WEIGHT:
19 ounces
MSRP:
$995
URL:
www.sigsilencers.com

We were concerned about what happens when the inevitable occurs and some asshat sends a projectile through your expensive, one-piece, hard-to-repair, Fed-involving Ti muffler, and it exits someplace other than where it's supposed to. We asked our friendly SIG rep about this, and he informed us that in his experience the most common place for a baffle strike was actually the end cap. In this case, SIG would simply cut off the cap, weld another in its place, and send the can back from whence it came without the owner needing to fill out another Form 4.

So there you have it. If you want a rifle suppressor that does it all (or as close to it as makes no difference), is competitively priced, and has that all-important Ti cool factor, check out the 338Ti from those nice chaps in New Hamster. We think you'll be glad you did. **R**

A CAN FOR ALL SEASONS

Advanced Armament's Ti-RANT 45M is the Leatherman tool of Pistol Silencers

By Mike Searson

Upon entering the world of NFA firearms, the typical firearm owner notices two things: It's expensive, and it can become an addiction. With machinegun prices soaring every year, the common thought process is "Get one while you can; they'll only increase in value." This is a result of an unconstitutional law being passed in 1986 that restricts current civilian ownership to those machineguns manufactured prior to May 1986.

In the realm of silencers, the expense is somewhat different. Silencers are still manufactured and can be owned by residents of 39 states. There's the cost of the silencer, which can range anywhere from $199 for a budget .22 LR model all the way up to $2,000 if you want to attempt to tame the sound signature on a rifle chambered in the mighty .338 Lapua or .50 BMG cartridges.

A smart shooter will perform a little research before purchasing a silencer, not unlike the same that might be performed when buying a vehicle or any piece of high-tech gear. A $250 silencer may sound like a bargain, until the shooter realizes that an extra $100 might buy one that is easier to clean, compatible with other calibers, or even quieter.

Dealer transfer costs (which can run anywhere from $50 to $250 per transfer) aside, purchasing a silencer represents a significant investment and as such, the can becomes a lifetime purchase.

Lastly, there's the matter of the $200 tax stamp. While many shooters claim it's the only tax they look forward to paying, we still feel the tax is unjust and like to minimize the amount of our shooting budget that goes into federal coffers.

If you're looking for that one pistol can to handle all your centerfire calibers you may want to look at the Advanced Armament Corporation (AAC) Ti-RANT 45M.

SIZE MATTERS NOT

The shorter Ti-RANT "S series" came about at the request of the

U.S. military. Unlike a civilian shooter who might prize the ultimate in sound suppression, the .Mil types wanted something short for a secondary weapon that might be riding in a holster. The handgun is generally a backup piece, and if deployed, that means something has more than likely gone wrong with the rifle. So if the bad guys in the immediate area already know your position, this would keep potential reinforcements from possibly being alerted.

The Ti-RANT 45M gives the best of both worlds in this regard — you can run it long or short and change its configuration in under a minute.

INTERCHANGEABILITY

Many people don't realize that certain .45 ACP silencers can be used on smaller calibers like 9mm, .40 S&W, .22 LR, or even 300 Blackout subsonic loads. AAC doesn't manufacture a piston to convert the latter, and shooting such a round will void the warranty if there's any damage — a supersonic 30-caliber rifle round will more than likely ruin the silencer, and yes, people do make mistakes and use the wrong round from time to time. However, it is possible to do it, just realize that the shooter assumes all risks.

The problem is that most silencers are threaded to a specific pattern, which means that the shooter needs to have the same pitch threaded to all his firearms on which he wants to run the silencer.

Or does it?

The Ti-RANT is based on the earlier AAC Evolution (which we looked at in our one-shot *ZEROED* edition from 2013). This means that the female threaded portion of the can rests in an interchangeable piston. On a can like the Evolution, the silencer was caliber specific and there were fewer than a handful of pistons available for either version. The Ti-RANT shares the same pistons across calibers, and if you have a multitude of firearms with differing calibers and male threaded ends, you can pretty much cover them all.

These pistons, known as the ASAP (Assured Semi-Automatic Performance) Clutch, form the base of the Nielsen device, which uses a spring in order to cycle the action on Browning-style pistols. When

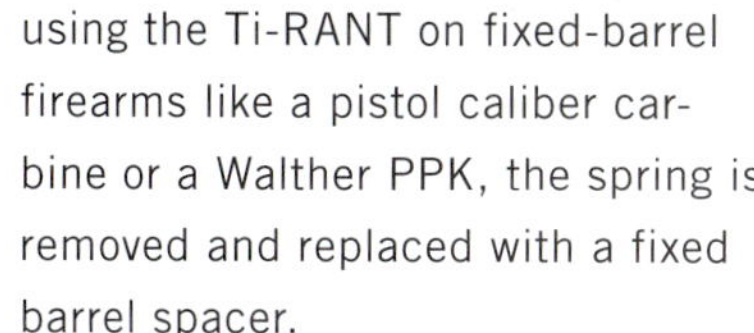

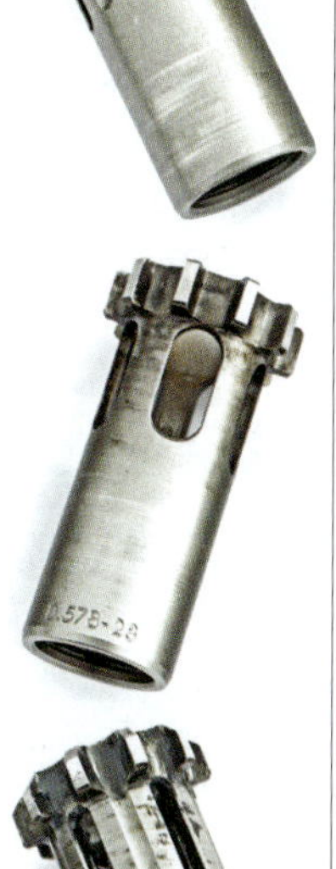

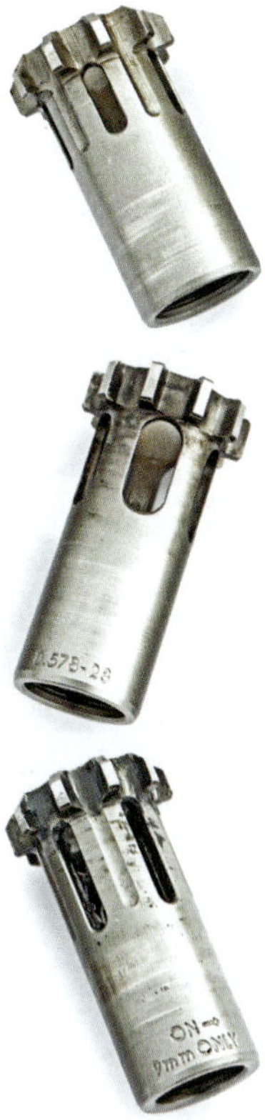

using the Ti-RANT on fixed-barrel firearms like a pistol caliber carbine or a Walther PPK, the spring is removed and replaced with a fixed barrel spacer.

The spring-loaded ASAP is housed in a steel blast baffle that takes the brunt of the initial blast and acts like a magnet for unburnt powder. More importantly, this booster takes the weight of the silencer off the barrel in order to allow the handgun to function properly.

As an added bonus, the shooter can tune the silencer to the pistol by pushing the silencer forward and rotating it in order to adjust the point of impact.

The ammunition provided for this review was courtesy of Freedom Munitions and their HUSH line, which is optimized for use with silencers as this goes beyond a simple subsonic factory loading.

The loads in the HUSH line minimize excessive unburnt powder. This results in a more effective silencer and a lot less carbon when it comes time to clean up.

TEST TIME

We ran the Ti-RANT 45M on every threaded barreled handgun in our safe. We started with the 45s and Freedom Munitions 230 Grain Hush subsonic rounds:

› HK USP Tactical 45

If there was ever a pistol designed for use with a silencer it's the USP 45 Tactical by HK. Wet, dry, with the short version or the full length: We find this to be the quietest handgun with any silencer. It may be the O-ring on the barrel providing slight drag on the slide or placement of the threads, but we suspect that the slide unlocks a microsecond later than everything else, keeping the gases from escaping through the ejection port.

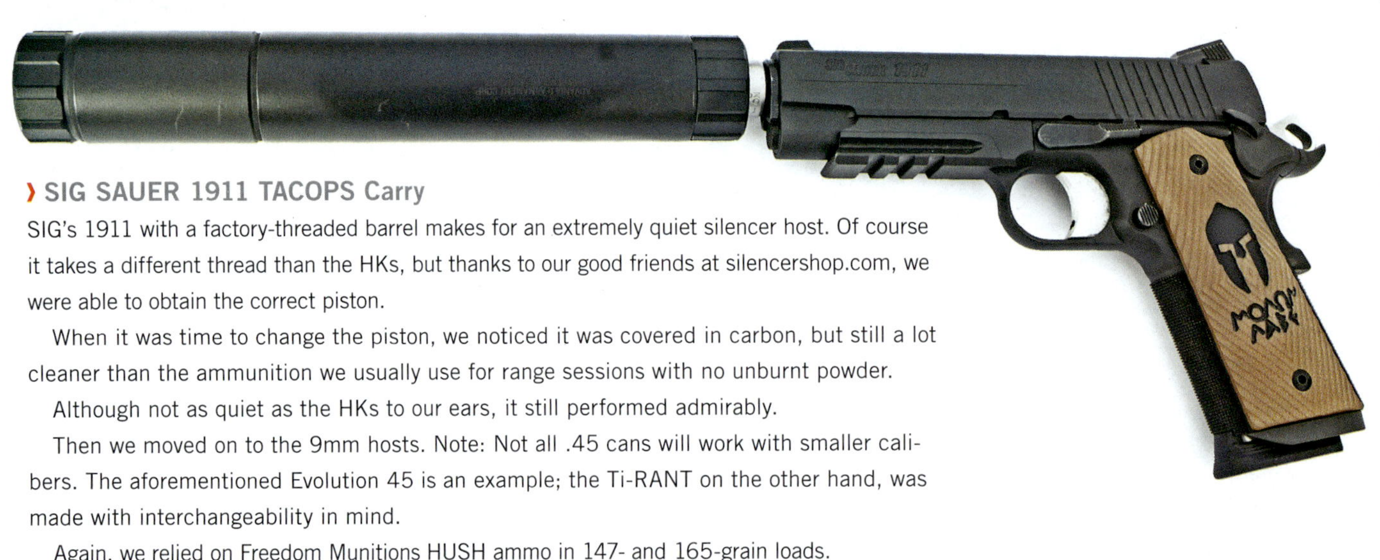

› SIG SAUER 1911 TACOPS Carry

SIG's 1911 with a factory-threaded barrel makes for an extremely quiet silencer host. Of course it takes a different thread than the HKs, but thanks to our good friends at silencershop.com, we were able to obtain the correct piston.

When it was time to change the piston, we noticed it was covered in carbon, but still a lot cleaner than the ammunition we usually use for range sessions with no unburnt powder.

Although not as quiet as the HKs to our ears, it still performed admirably.

Then we moved on to the 9mm hosts. Note: Not all .45 cans will work with smaller calibers. The aforementioned Evolution 45 is an example; the Ti-RANT on the other hand, was made with interchangeability in mind.

Again, we relied on Freedom Munitions HUSH ammo in 147- and 165-grain loads.

› HK USP Compact Tactical 45

Like its slightly bigger brother, the USP CT was a joy to shoot. Our host pistol relies on extra tall Heine "Straight 8" sights, which we preferred to the factory USP sights. When we can see over the can, we won't say "no."

› SIG SAUER P229

The P229 has a slow lockup time by a few milliseconds and results in a quiet shooter. We were particularly impressed with the sound on the shorter version of the Ti-RANT with this pistol.

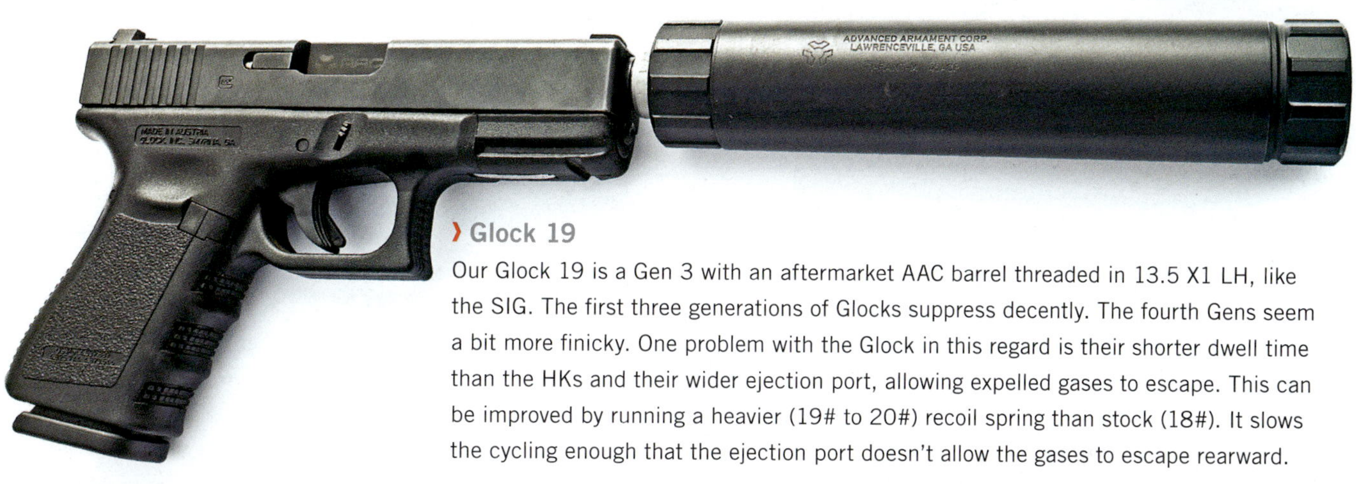

› Glock 19

Our Glock 19 is a Gen 3 with an aftermarket AAC barrel threaded in 13.5 X1 LH, like the SIG. The first three generations of Glocks suppress decently. The fourth Gens seem a bit more finicky. One problem with the Glock in this regard is their shorter dwell time than the HKs and their wider ejection port, allowing expelled gases to escape. This can be improved by running a heavier (19# to 20#) recoil spring than stock (18#). It slows the cycling enough that the ejection port doesn't allow the gases to escape rearward.

› WHAT'S IN A NAME?

"Silencer," believe it or not, is the correct term and was coined by Hiram Percy Maxim in 1906 for his invention to reduce firearm noise. To date this is the legal term used by the BATFE and the overwhelming majority of manufacturers who build them. "Suppressor" or "muffler" may sound like a more accurate term with regard to the action being performed, but it is really a form of slang just like "can" or "hush puppy."

› SHOOTING WITH THE SILENCER

Some shooters prefer taller "Suppressor Sights" (Yeah, yeah, we know what we said, but that's what the pistol manufacturers like to call them) and some might opt for an electronic sight like the Trijicon RMR to see over the silencer. However, there's another method, which may not be taught by every "Certified Instructor" known as "Shooting through the can." The front sight is still the primary focal point, only in this instance the silencer is superimposed over the target.

It takes a little getting used to and can be improved upon by using contrasting sights such as an orange front and white or green rears. In the case of the Beretta, we paint the rear of the front sight orange.

› NIELSEN RATINGS

The original Nielsen Device was a modular component that had to be added to the suppressor or the host pistol and actually hampered sound reduction, making the suppressor slightly louder.

It was designed by the late Charles "Mickey" Finn at the request of a fledgling Naval Special Warfare Unit whose name rhymes with "SEAL Team Six" for use with Finn's Qual-A-Tec suppressors that the SEALs were using on their pistols.

When going from the M9 (Beretta 92), which did not need the device, to the SIG P226 or MK23 SOCCOM pistol, there were reliability issues that prompted its invention.

AAC chose to make the Nielsen Device an integral part of their centerfire pistol cans and, in doing so, realized they could make interchangeable rear caps and pistons so the suppressor could be used on more than one pistol with different thread patterns such as ½-28, ½-32, and Metric 13.5 x 1 LH.

The namesake of the Nielsen Device was the maiden name of a family member of Finn's.

› THE MAKING OF A TI-RANT

Despite its name the can isn't entirely constructed of titanium. Only the outer tube gets that space-age honor and the Grade 9 titanium offers improved strength coupled with a significant weight reduction. The baffles are made from hard coat anodized 7075-T6 aluminum alloy.

The 45 M differs from the fullsized and S models in that the first 3 inches of baffle stack are removable and can be replaced by the endcap to transform into a shorter silencer.

› HUSH AMMO

Most shooters running suppressors typically run the most convenient subsonic ammunition that they can find, which in the case of the .45 ACP is 230-grain loads and in 9mm 147-grain or heavier.

Run-of-the-mill factory loadings can often throw unburnt powder into the can or the shooter's face, even if it's branded as "subsonic."

Freedom Munitions HUSH line was developed specifically for use with silencers, and from the rounds we tested we experienced little of these problems. The result was a quieter, more efficient silencer that was easy to take apart and clean.

› FIRST ROUND POP

The first round through a silencer at a range session is often the loudest. The root cause for this is the silencer is full of cooler air containing oxygen that reacts with hot gases from the fired round. After that first shot, the oxygen is depleted and subsequent shots are quieter.

› WHERE TO GET ONE?

The Ti-RANT 45M is shipping regularly now; most NFA dealers can order them for you if they're not in stock. If you're brand new to the process and need a trust or the lowdown on finding a local dealer, you can order from silencershop.com. They're the leading distributor for silencers and their unique ordering system makes it a hassle-free transaction, plus they spend a significant amount of time and energy helping to reform NFA laws in every state. If you live in one of the states that just decriminalized silencers and have no idea where to find a dealer, check them out.

› A WET CAN?

Pistol silencers can benefit from a slightly reduced sound signature by running them with about 5 cc of water (a bottle cap full). Simply pour the water into the rear of the can, hold your finger over the hole in the front and give it a quick shake. The water acts like an ablative and cools the hot gases eliminating the dreaded "first round pop" (see above).

Despite some people advocating wire pulling gels, WD-40, or hippie tears, the experts and manufacturers advocate using water above all else.

› Beretta 92

We expected the Beretta to sound louder than it did due to the open slide design and were pleasantly surprised with its performance. We had two issues with the rear endcap impacting the plastic guide rod and causing a stoppage. This was in part our fault for threading an existing factory barrel. Aftermarket barrels tend to be a bit longer and don't have this issue.

There is some debate about using a fixed barrel spacer on the Beretta — AAC told us it wasn't necessary, but other silencer manufacturers do advise running them on the M92. Using the ASAP won't damage the suppressor, but it may help if using an aftermarket threaded barrel with a longer thread pattern.

› Steyr M9A1

There's another Austrian pistol designer, more famous for their rifles, known as Steyr Arms. The S9A1 subcompact model is this author's regular CCW pistol, so we wanted to see how its parent design fared with a can.

The high grip angle coupled with a low bore axis makes for an accurate shooting platform, and the trapezoidal sights make "shooting through the can" an easy feat as the front sight is a hi-vis triangle.

What's interesting about the M9A1 and the Ti-RANT is that this pistol can be a fickle suppressor host. We've tried this pistol with dedicated 9mm silencers and a variety of ammunition with less than acceptable results.

However, we experienced no such stoppages with the Ti-RANT 45M. This may be due to the larger volume of the silencer and a slight boost in recoil.

› S&W M&P

Along with .45 ACP and 9mm, the Ti-RANT is rated for .40 S&W. Our host gun was a Smith & Wesson M&P with an aftermarket threaded barrel and, of course, the appropriate piston.

This caliber is not ideal for suppressed shooting as the barrel threads seem to be a weak point in most instances due to the fact that many firearms in this caliber were designed for 9mm in mind and opening up that barrel by 1 mm doesn't leave a lot of meat for threads to be properly cut for long-time use.

We didn't use Freedom Munitions ammo in this handgun; instead we relied on some subsonic 180-grain hand loads we developed ourselves.

The round performed well according to the sound meter, but we thought it sounded the loudest of everything we fired during this test.

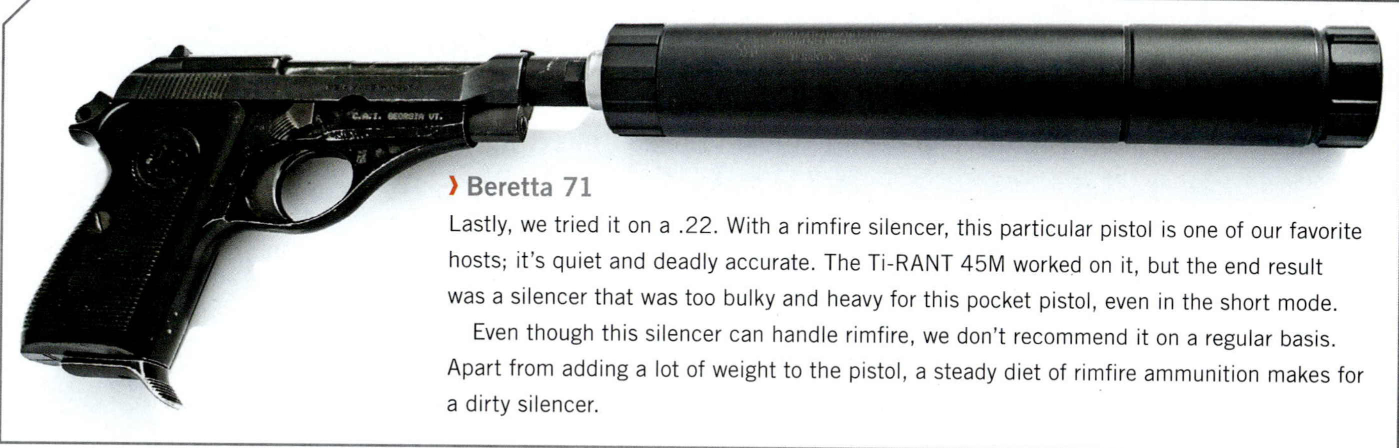

› Beretta 71

Lastly, we tried it on a .22. With a rimfire silencer, this particular pistol is one of our favorite hosts; it's quiet and deadly accurate. The Ti-RANT 45M worked on it, but the end result was a silencer that was too bulky and heavy for this pocket pistol, even in the short mode.

Even though this silencer can handle rimfire, we don't recommend it on a regular basis. Apart from adding a lot of weight to the pistol, a steady diet of rimfire ammunition makes for a dirty silencer.

› Other Calibers

The Ti-RANT 45, 45M, and 45S are fully capable of working with calibers such as .44 Special, .45 Auto Rim, .45 Colt, .38 Special, and anything else with a bore diameter less than 0.455 inch and a muzzle velocity of less than 1,080 feet per second.

› .300 Blackout?

Based on our experience with the silencer and others of a similar design, we believe that it would work well with 300 Blackout subsonic loads and a fixed barrel spacer in lieu of the spring in the ASAP. Unfortunately AAC does not offer an adapter for this purpose. This is probably a good thing as accidentally firing a supersonic load through the can would have disastrous consequences.

RESULTS BY THE NUMBERS

Here are the tabular results to see how each pistol performed. Each rating is an average (eliminating first round pop) of 20 rounds fired from each pistol in each configuration.

On the .45-caliber pistols and 9mm pistols, the can performed exactly as it was supposed to. We tried some 9mm out of another 9mm silencer (AAC Illusion 9) and while it metered quieter, we found the Ti-RANT to deliver a more pleasant sounding tone.

MAKE: AAC
MODEL: Ti-RANT 45M
LENGTH: 8.74 inches
DIAMETER: 1.38 inches
MSRP: $850
URL: www.advanced-armament.com

Shooting with a silencer goes beyond mere decibel reduction. Different handguns in the same caliber will sound different due to the difference in milliseconds between the action unlocking, the length of the barrel, the barometric air pressure, altitude, and where you happen to be shooting.

Silencers at indoor ranges sound amplified due to echoes; shooting suppressed outdoors in a wooded area can sound similar. Shooting in the desert at 5,000-plus elevation and no humidity makes everything sound much quieter.

This may seem as if we're deviating from "the science," but we're not.

Sound is a vibration that begins as an audible mechanical wave of pressure and displacement, through a medium such as air. How we hear sounds is a matter of perception by the brain. This is further impacted by the shooter's hearing sensitivity.

Look at the diversity in music tastes among your friends — beyond style or genre, certain notes or tones have a different appeal to everyone. It's the same with silencers. We provide decibel readings because it's somewhat measurable from a manufacturer's standpoint, but it can be like lumen ratings on flash lights and only tells a small part of the story.

The interchangeability of calibers and ability to run short or long gives the Ti-RANT 45M an edge, making it a necessity if you have a diverse assortment of threaded pistols you want to quiet down and don't want to send off more than one $200 check to the Treasury Department for another stamp.

More importantly, with the exception of .40 S&W through the can, it's music to our ears. R

DECIBELS

HOST	CALIBER	LONG (DRY)	LONG (WET)	SHORT (DRY)	SHORT (WET)
HK USP 45 Tactical	45 ACP	133	126	139	134
HK USP 45 Compact Tactical	45 ACP	134	127	140	134
SIG SAUER Tac Ops Carry 1911	45 ACP	136	130	141	137
SIG SAUER P229	9mm	129	124	133	129
Glock 19	9mm	130	123	134	131
Beretta M92	9mm	130	125	136	132
Steyr M9A1	9mm	129	124	132	129
S&W M&P 40	.40 S&W	135	131	145	138
Beretta M71	.22 LR	127	125	129	128

SOVI3T STYLING, AMERICAN EXECUTION

We Go Hands on with Dead Air and Gemtech AK Cans

By Dave Merrill and Iain Harrison

The *idea* of slapping a suppressor on an AK has always been appealing. In practice, however, an AK with a silencer can leave much to be desired. Right from the start you're fighting an uphill battle. There are several issues at play, each being a potential deal breaker in their own right. And when they form a confederation? Let's just say that after a couple of attempts ourselves, we haven't bothered to try again in a long time.

Part of what gifts the mythological reliability of the Kalashnikov is the shear volume of gas the system puts out. Sticking a muffler on the muzzle can surge the gas flow, causing issues. Not only is recoil more violent, but the rifle is louder than other rifles at the shooter's ear due to increased ejection port exhaust. You're also amping up the wear and tear on your rifle.

Muzzle thread concentricity is an ongoing concern with AK rifles. Having bore and threads aligned to NASA-level tolerances is less of a concern when running a short, slant brake. But, mount 7 or 8 inches of suppressor on the end — and concentricity matters a helluva lot more. Soviet manufacturing was never highly regarded for its precision, and that theme continues with many an AK, despite advances in modern manufacturing techniques. Furthermore, mounting solutions for commie thread patterns are nowhere near as common as their western brethren.

Today we'll be covering two different, but equally brand-new (and all-American) options from Dead Air and Gemtech. Both companies approach the unique challenges of putting a can on a Kalashnikov from discrepant directions, but they have the same goal dead in their sights.

While there are umpteen methods to address these issues such as cutting new threads and having other [invariably expensive] custom work done, Dead Air set out to solve them all in one go. Instead of trying to tailor the rifle to the silencer, they decided to tailor the silencer to the rifle.

They call it the Wolverine PBS-1. As the name implies, it's aesthetically patterned after the original PBS-1. We've seen cans like this before from some other manufacturers, but unfortunately a few of those companies stayed a little too true to the original, unexceptional Soviet design with layouts largely consisting of thin and straight baffles. They typically had the same unexceptional performance, too.

Dead Air promised the internals would be modern, so we headed out to Utah to meet the Wolverine PBS-1 in person. Bear in mind, our observations are based on preproduction samples. Some aspects of the Wolverine PBS-1 may change by the time it ships. We'll point out anticipated changes as we move along.

In order to address the gas issues, the team at Dead Air used their Sandman-S silencer as a starting point. The Sandman-S is a lower backpressure model, which is louder at the muzzle than the larger Sandman-L model, but is quieter at the shooter's ear.

"It's a balance," says Mike Pappas, HMFIC at Dead Air. "If you maximize muzzle suppression, you pay the price at the ejection port."

Internally, the Wolverine PBS-1 is not exactly the same as the Sandman-S. It's evolved into its own design, but the team at Dead Air retained the desired lower backpressure. And, its endcaps are interchangeable with the endcaps from the Sandman-S, though the final standard exit diameter had yet to be determined.

The suppressor core is machined from Stellite and fully circumference welded for maximum strength. "We wanted to make it as tough as the rifle it's made for," says Gary Hughes, AK nerd supreme of Dead Air.

Not only did the PBS-1 styling allow for a classic look and feel, but that bulbous base allows for a multitude of threading arrangements. While the Wolverine PBS-1 is a direct thread silencer, there's a little more going on than initially meets the eye. The mounting system is a two-piece affair consisting of a thread adapter (available in a number of foreign and western pitches) and a locking collar.

The adapter and locking collar are counter threaded to each other to prevent unintentional removal. In other words, if the thread adapter is a left-hand thread, the locking collar is right-hand. To install, the Wolverine is first threaded onto the barrel as far as it will go, and then the locking collar is tightened down to the rifle. There are relief cuts in the locking collar to engage the spring-loaded pins on AK front sight bases to further fasten the suppressor and keep everything in place.

On the preproduction models we shot, it took a little bit of work to lock everything down. It wasn't difficult, but it took a couple of tries to get comfortable with the mounting system. Dead Air says this is one of the areas they're looking to improve on production models.

Due to the, ummm "generous" tolerances of foreign thread patterns, the bore diameter of the Wol-

Dead Air Sandman-S (left), Wolverine PBS-1 (right)

verine PBS-1 is slightly larger than your average silencer. One of the reasons for a replaceable endcap is that even with the most egregious threading, the endcap is still the part most likely to be struck by an exiting projectile. Dead Air says this will be an exceptional performer for short-barreled rifles, even on non-AK western guns. For us, we plan to mount one on a SCAR CQC for maximum annoyance and confusion.

All of this adds up to: Don't do anything special, just put it on and shoot.

SO HOW'S IT SOUND?

We don't hold an OSHA compliance verification rating, but nobody shooting the Wolverine complained about the noise levels. We were comfortable shooting without hearing protection in the open air. The general recommendation for any supersonic suppressed centerfire is going to be to wear some ear pro. No issues with short-barreled 7.62x39 nor full auto, or even when mounted on a rare Yugoslavian 7.62N AK.

Pick one up and in no time you'll be speaking with a terrible fake Russian accent and toasting *За здоровье!*

Anticipated release is late summer 2016 with an MSRP of $1,050.

The second of our dedicated AK cans this issue hails from the great state of Idaho, where Gemtech created it in a joint venture with Arsenal. The brand-new, as of March 2016, design was developed in order to complement Arsenal's lineup of both 7.62 and 5.45 rifles. That means this is a 100-percent titanium unit that adds only 19.5 ounces and 7.5 inches to the muzzle, despite being full-auto rated with a heavy-duty blast chamber. Although it's presently available only as a package deal through Arsenal, there are plans to release it as a separate product in the future.

MAKE:
Dead Air Armament

MODEL:
Wolverine PBS-1

CALIBER:
7.62N and below, to include 5.56, 5.45, and 7.62x39

CONSTRUCTION:
Stellite and Stainless Steel

LENGTH:
7.2 inches

DIAMETER:
1.5 inches (body), 1.9-inch (base)

WEIGHT:
20.8 ounces

DECIBEL READINGS:
137dB [muzzle] 142dB [at-ear] (from 16-inch 7.62x39 AKM)

MSRP:
$1,050

URL:
www.deadairsilencers.com

In order to ensure longevity in a platform that's notorious for third-world levels of quality control, Arsenal hand-selects each rifle that's to be paired with a suppressor, checking for concentricity and runout in both the bore and muzzle threads. A proprietary, two-lug Quickmount adapter is then installed, which also serves as a flash suppressor when the can is not in place.

We shot the suppressor on both 7.62 and 5.45 rifles and had little in the way of expectations when it came to being hearing safe. Although we didn't have a dB meter on hand, report from the supersonic, steel-cased ammo was comfortable in the 50-yard, U-shaped bay we used and the perceived noise of bullets on steel was louder than the muzzle blast.

Again, this was a purely subjective test, but we wound up putting several magazines of each caliber through the same can, with the small-bore sounding quieter than its big brother. Swapping between guns proved to be an easy (if hot) exercise, which we expect has something to do with the can reciprocating about 1/8th of an inch on the mount at each shot, as shown by our high-speed camera. This movement may serve to scrape carbon from between the mount and the corresponding surfaces in the suppressor body. **R**

MAKE:
Gemtech

MODEL:
Arsenal, Inc. AK

CALIBER:
30 cal

CONSTRUCTION:
Titanium

LENGTH:
7.5 inches

DIAMETER:
1.625 with shroud

WEIGHT:
19.5 ounces

DECIBEL READINGS:
7.62x39 / 142.4 at 1 meter left / 138.5 at shooter's ear
5x45x39 / 142.4 at 1 meter left / 138.0 at shooter's ear

MSRP:
$999

URL:
www.gemtech.com

ROAD TO DAMASCUS

A Modern Spin on an Ancient Art

By Iain Harrison
Photos by Stickman and Iain Harrison

Shrouded in mystery, Damascus steel has an aura surrounding it like no other material. According to lore, the blade of a Damascus sword can cut cleanly through a gun barrel and will sever a hair falling on it. Because the knowledge needed to produce this ancient material was lost in the early 1800s, it's achieved a mythical status. While we can approximate it in modern times, the ancients took their secret recipes and techniques with them to the grave.

A true Damascus blade utilizes a type of crucible steel originally produced in India from around 200 B.C., comprising a Martensite or Pearlite matrix, within which are micro carbide layers that produce microscopic "teeth" in a cutting edge. In recent years, researchers have also discovered evidence of carbon nanotubes and tungsten micro alloys, adding to the mystique. This Wootz steel features a distinctive pattern, which, depending on the individual sample, can look like waves or flowing water. It was traded across the Near East for almost two millennia, much of it fetching up in the city of Damascus (which had a thriving arms industry), where it was forged into the blades bearing its name.

MAKE:
Nottingham Tactical
MODEL:
Mokuti Suppressor
CALIBER:
30
CONSTRUCTION:
100% Mokuti titanium laminate
LENGTH:
8 inches
DIAMETER:
1.5 inches
WEIGHT:
16 ounces
URL:
nottinghamtactical.com

MAKE:
Dahmer Arms
MODEL:
Damascus Lower Receiver
CONSTRUCTION:
Damascus steel
WEIGHT:
28 ounces
URL:
www.dahmerarms.com

While original, genuine Damascus steel is no longer available, the Damascus look is still highly sought after and can be visually reproduced by another technique known as pattern welding. In it, different alloys are heated and then forge welded (think BFH) together before being folded over and forge welded again. This produces a laminated material that has an amalgam of the characteristics of the materials that went into its creation.

It is this method that's used to produce the two items here. Be warned, if you admire the flowing, organic lines of either original Damascus steel or its modern, forge welded counterpart, you'll need a healthy bank account if you plan on adding it to the collection.

NOTTINGHAM TACTICAL MOKUTI SUPPRESSOR

The Japanese art of Mokume-game was originally applied to sword making, where it would be employed in the creation of eye-catching tsuba or other decorative components. The Mokuti billet used in Nottingham Tactical's suppressor was created in an inert atmosphere under intense heat and pressure by Chad Nichols in Blue Springs, Mississippi. If that sounds like an expensive way to create a chunk of titanium bar stock, you'd be absolutely correct. Before attempting to machine this piece, (and machining it is a royal PITA) the maker was already 10 grand in the hole due to material costs, so the finished price tag of somewhere north of $14,000 seems almost reasonable. There will only be five of these cans ever produced, so take a number and get in line.

Using modified K baffles, each of which is also spun from the same material as the endcaps and tube, the can is rated for .300 Win Mag and is just the thing for your next unicorn hunt.

DAHMER ARMS DAMASCUS AR LOWER

CNC'd from a billet comprised of 416 layers of carbon steel, the Dahmer Arms lower is for those who want an AR unlike any other. Or, rather, unlike any apart from the other nine units that make up this production run. For those interested in the technical aspects, the alloys used were 203e, 15n20, 52100, and 5160 and the billet was folded by hand on a Nazel 4B power hammer (if you're interested in what one of these mechanical behemoths looks like, check out RECOIL's interview of Jesse James in Issue 11).

Once off the machines, the receiver was etched to bring out the grain pattern and sealed against rust. So there you have it — pair your billet Damascus lower with a Mokuti can and you'd have a one-of-a-kind rifle that'll pull envious stares wherever you go, or, for the same kind of money, you could have a pretty decent family sedan. We know which one would be more memorable. **R**

STINGERWORX EMPEROR SERIES SUPPRESSOR

The Emperor Does Have Clothes

By Iain Harrison

Photos by Kenda Lenseigne

At RECOIL, we're always on the lookout for product innovations from the little guy. Let's face it — if all we ever wrote about were the top-selling guns from industry leaders, you'd get pretty tired of reading Glock 19 articles. Likewise with suppressors. Sure, SIG and Silencerco have great and extensive product lines, but there are many ways to skin a cat, some of which are developed in garages and small shops all over this great land.

Stingerworx is a small startup company with big ideas, and we went to the range with Dave Anderson, its founder and chief engineer, for a look at what they've been working on.

EMPEROR SERIES SUPPRESSOR

Due to our country's onerous and utterly illogical suppressor laws, there's a very real benefit to cans that fill as many roles as possible. With the ATF taking up to a year to process a simple transfer and the attendant $200 fee, we want our silencers to work on as many different guns as their materials and engineering will permit. Which is why there's a tendency to overbuild everything here — quite the opposite of countries in which they're treated the same way as the firearms they're mounted upon.

Even in the U.K., which has some of the most draconian infringements of civil liberties imaginable, you can get a $35 disposable .22 suppressor or an 11-ounce carbon-fiber centerfire can for less than five Benjamins. For U.S. shooters who need the same suppressor to work on a 24-inch barreled .308 hunting rifle as well as a 7.5-inch barreled SBR, there's going to be a significant weight penalty imposed on the resulting design.

Manufacturers attempt to address some of the issues involved in creating a can for all seasons by offering modular suppressors that can be tailored to the job at hand. We covered Rugged Suppressors' offerings in RECOIL Issue 23, and the grandly titled Emperor is Stingerworx' solution to the same problem. Instead of adding or subtracting additional baffles to the front of the stack, the Emperor adds them to the rear, in the form of a removable over-barrel chamber. Reflex designs aren't anything new, but the combination of increased chamber volume and a novel blast baffle design

STINGER - DEFENSE
NAMPA, ID
EMPEROR 762-L2
7.62MM 0000172
PATENT PENDING

derived using gas-flow dynamics modeling add up to a significant reduction in noise. How much? According to Anderson, they're seeing an additional 2 to 3dB reduction, depending on caliber.

There's a perennial debate among suppressor nerds, along the lines of Glock versus 1911 or 9mm versus .45, centering on the question of whether rifle cans should be welded or threaded together. There are strong arguments on both sides; Stingerworx went the threaded route. "We wanted the ability to tailor the suppressor for its intended use and use different materials where it made the most sense," Anderson says.

While the standard Emperor uses titanium throughout, there's nothing preventing the addition of an Inconel blast baffle (at a cost of additional weight) for use on particularly ablative guns — this is very tricky to pull off in a welded design, as anyone who's tried sticking together metals from different parts of the periodic table will attest.

It also means that in the event of a baffle strike, the end cap can be quickly unscrewed at the factory and it, or the rest of the stack, replaced at no cost to the customer. Of course, it would be even more convenient if we as end users could be trusted to have spare parts on hand, but according to the regulations as currently interpreted by unaccountable bureaucrats, that would land you a stretch in Club Fed.

MAKE:
Stingerworx

MODEL:
Emperor Series 30-L2

CALIBER:
Up to .300 Win Mag

MATERIAL:
Titanium

DIAMETER:
1.75 inches

WEIGHT:
13.1 ounces; 17 ounces (with additional blast chamber)

MSRP:
$975

URL:
www.stingerworx.com

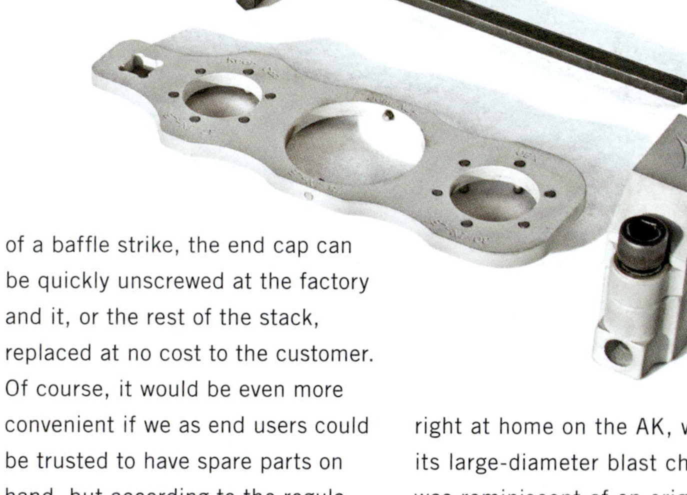

We shot the 30-L2 on a variety of firearms, from a JP Rifles CTR1 in 6.5 Creedmoor to a Century Arms AK, and while we don't have access to sound measuring equipment, the end result was pleasing to our uncalibrated ear. It looked right at home on the AK, where its large-diameter blast chamber was reminiscent of an original Soviet PBS-1, receiving admiring glances from fellow shooters on an adjacent bay.

However, perhaps the most remarkable feature of the Emperor isn't its looks, but rather its weight or lack thereof. At only 13 ounces for a full-auto-rated, multi-caliber suppressor, it's a noteworthy addition to a crowded field. **R**

RIMFIRE RATED

RECOIL Compares 10 Rimfire Cans

By Rob Curtis

Break out the plinkers, boys. 22LR is back on the menu!

Some of you will grumble "There's no 22LR at *my* Walmart," but national stocks seem to be replenishing and it's reportedly even on the shelves in the deepest corners of liberal New England and Southern California. Sure, it's not nearly as cheap as it once was, but it's still cheaper than 9mm.

What's better than shooting an easy handling, light-recoiling 22? Shooting a quiet one. If you don't have a can on your plinker, you're missing out. We compared an array of rimfire cans you might be interested in as you put your favorite deuce-deuce back into rotation.

THE CORE

Suppressors work by disrupting and slowing down the flow of exhaust gases following the bullet out of the muzzle.

There are two overarching categories of suppressor design — those that use stacked individual baffles and those that use a monolithic core, commonly called a monocore.

Monocores are generally made from aluminum, instead of steel or titanium, because it's easier (read: cheaper) to machine and it's lighter. Monocores are also pretty easy to remove, clean, and reassemble since the guts are just one big piece of metal. The aluminum isn't as durable as steel, though, so yanking a really baked-in core from its tube can be exciting. Some aluminum cores are anodized, increasing the monocore's durability, though scratching or gouging the finish during cleaning will lead to accelerated wear.

Stacked baffles are generally made from steel, though there are aluminum-baffled cans out there. Suppressor cores made from stacked baffles generally enjoy improved sound reduction performance compared to monocores because individual baffles can be machined more intricately, and therefore tuned more effectively for a given purpose. They're also

SOUND REDUCTION PERFORMANCE AND ACCURACY								
MAKE	MODEL	AVG DB, PISTOL /SUPER	AVG DB, RIFLE/SUPER	AVG DB, RIFLE/SUB	OVERALL AVG DB	FRP	GROUP MOA (SUPERS)	POI SHIFT (MOA)
Dead Air Armament	Mask-HD	122.4	119.9	114.6	118.9	3.3%	2.7	1.90
Gemtech	GM-22	125.8	119.2	117.7	120.9	7.7%	2.8	0.60
Gemtech	MIST-22	N/A	124.3	117.9	121.1	0.0%	1.2	N/A
Griffin Armament	Checkmate QD	127.4	119.4	116.0	120.9	0.0%	1.8	4.20*
Q	El Camino	122.8	121.5	118.0	120.8	3.9%	2.2	2.80
Ruger	Silent-SR	122.8	123.4	117.8	121.3	3.1%	1.9	0.60
Sig Sauer	SRD22X	117.7	121.3	116.6	118.5	7.1%	2.3	1.50
SilencerCo	Osprey Micro	122.6	120.7	116.9	120.1	4.1%	2.2	1.90
SureFire	Ryder 22-S	123.4	121.8	117.2	120.8	4.8%	2.6	1.50
Thunder Beast Arms	22 Take Down	118.0	119.5	116.8	118.1	4.9%	2	1.00

*Defective mount

tougher and will withstand more aggressive cleaning methods.

Baffle designs are the secret sauce in any can. K-baffles, M-baffles, cone baffles, slant baffles, pig-nose baffles ... these all describe designs attempting to disrupt the flow of gases surrounding a projectile as it leaves the muzzle of the rifled barrel without affecting the bullet's flight path.

In some cans, the first and last baffle in a stack, otherwise called the blast and normalization chamber, are dimensionally different than the middle baffles. In these designs, manufacturers have begun numbering the baffles to make reassembly faster and foolproof. Some baffle designs require each primary baffle (the middle baffles) in the stack be timed to prevent lopsided cross-jetting that affects accuracy. These have tabs and slots that mate up for foolproof reassembly.

BASELINE PERFORMANCE

The cans we tested were all quiet. Some were quieter than others, but unless shooting them back-to-back-to-back, it's hard to tell them apart based on absolute noise level alone. That's to say, rimfire can design has gotten to the point that we'd have to work to find an objectionably loud one.

A caveat to this observation comes with first round pop (FRP). Oxygen in the can fuels a louder pop on the first round. Subsequent rounds are quieter as oxygen is displaced by inflammable gases produced by cartridge ignition. Some cans have startling FRP, some have none. FRP will vary with ammo selection, weather, and proximity to campaigning politicians. We included the FRP measurement in our readings as a percentage increase over the average sound of the can.

Accuracy, handling, FRP, and features such as quick attach, easy disassembly for cleaning, and cartridge versatility are what set competitors apart. We'll take a can that's 1 dB louder than the next one if it's more accurate, runs longer between cleanings, and we can break down without hunting for a proprietary disassembly tool.

Since we aren't mob hitmen, pressing muzzle to temple in a back alley, accuracy outweighs stealth for us. So, when we chose our favorites, we weighted accuracy measurements a little more heavily than the other considerations.

CENTRAL CONSIDERATIONS

Sound: There's no one quietest can no matter what the advertising copy says. The quietest can on a rifle might be a poor performer on a pistol. And when looking at sound pressure testing numbers, consider them perishable. Is the manufacturer's dB specification taken at sea level or in the mountains? It matters. Data is only good for comparison when taken on the same day with the same ammo shot from the same gun. Comparing these 10 cans with our data works. Combining or comparing our data with a manufacturer or another review won't.

Further, sound reduction performance on one platform doesn't equate to the same performance on another. For example, cans made for and used on rifles are generally quieter than pistol cans in this application, because their design accounts for exhaust gases that have already been slowed and cooled by the longer barrel. The same can on a pistol may sound like a cannon. Ammo also plays a role in sound output.

Comparing first round pop numbers only tells part of the story, too. Cans with straight or simple baffle channels refill with oxygen quicker than others, bringing FRP back at the beginning of every string.

Check out our numbers, and you'll see you can get a can that works with the particular firearm you're likely to use, or you can choose one with the best average performance across platforms and ammo types. But you'll be hard pressed to find one can that rules them all.

Weight: Hanging a can on the end of a barrel, essentially a lever, magnifies its weight. Every ounce contributes to shooter fatigue and can impact accuracy.

Size: Along with weight, the size of a can affects the handling of the host firearm.

Accuracy: Porting between baffles and the bore size of a can determines how much turbulence a bullet passes through on its way through and out of the sup-

pressor. If turbulence is managed effectively, accuracy is unaffected, and in some cases it can be improved. We want a can that maintains the same point of impact when the can is taken off and put back on the gun. While POI shift ranged from 1 to 2.8 MOA among the contestants, we found repeatability on all the cans was excellent. The Griffin Checkmate QD had some issues here; we'll address those in the notes below.

Ease of Maintenance: Nobody likes pulling out the manual when you can't remember how something's supposed to be oriented or tearing the tool bench apart looking for a special endcap wrench you now remember you left at the range. We gave the highest ratings to the least complicated, most forgiving cans. We also like cans with captured tubes that turn the muzzle mount instead of twisting off in your hand when removing the silencers from the host.

Versatility: We gave high marks to the cans with the widest latitude of ammunition types and configurations. A full-auto fire rating says the manufacturer built its can tough enough to deal with full-auto 22LR. That means it'll withstand the harshest 22LR firing schedule you and your 12-year-old can come up with on a lazy Sunday in the backyard.

Price: We rated each can's price based on the prices of its competitors. But, we think value is a more important measure, and that's reflected in our editor's recommendations.

TESTING

We tested the cans on a pair of ubiquitous 22LR platforms: the Ruger 22/45 Mark III pistol and the Ruger 10/22 Tactical rifle (model 1261). We logged sound readings with a Larson Davis sound meter from both platforms and performed accuracy testing with the 10/22 at 50 yards. All the sound metering was done over the course of a couple hours. (Environmentals: 50 degrees F, 29.26 inches barometric pressure, 144 feet above sea level, 40 degrees F dew point, and 84 percent humidity.)

The firing schedule for each can consisted of an initial burst of 50 pistol rounds to seal the cans up before taking sound measurements. Once we got sound data, we plugged the cans back into the pistol for 200 rounds, each. Then we moved the cans to the 10/22 and shot groups. With accuracy data logged, we switched back to the pistol and filled the cans with 250 more rounds of rapid fire to get them dirty and see how much of a fight they'd put up at cleaning time. (Meanwhile, we're hoping the embattled EPA doesn't have the resources to declare a tree stump in our backyard a micro Superfund site after we pumped 5,000 rounds into it.)

For ammo, we used a mix of Gemtech 42-grain Silencer Subsonic and Federal bulk pack 40-grain Target Grade Performance. The upside to the monotonous 5,000-round celebration of lead and carbon was the opportunity to note the reliability of the ammo. The guns choked on about three per 1,000 of the Gemtech rounds, while the Federal bulk pack averaged 20 clicks per 1,000.

The first 250 rounds through each can were Federal, the second 250 were Gemtech. We metered the Federal on both rifle and pistol, but only metered the Gemtech subs on the rifle cans, because who shoots subs from a pistol?

Our thumbs compel us to tell you about a couple of loaders we used. Stuffing 10/22 mags with a Butler Creek 50 Round Hot Lips Loader 10/22 ($25 to $30 on the street) was much better than thumbing them into those BX25 mags. Rarely, a round got stuck in the works, but clearing it was a matter of pulling the mag and shaking the thing. (We also tried a Champion Target 10/22 loader that was a PIA all the way around.)

For the Mark III mags, we used a McFadden Ultimate Ruger MK Clip Loader ($18 on street). As long as we were careful to keep the rounds level as they feed, it worked flawlessly.

DEAD AIR ARMAMENT MASK-HD

MSRP
$449

URL
www.deadairsilencers.com

CALIBERS
.22LR, .22WMR, .17HMR, 5.7x28mm

CONSTRUCTION
Compressed K Baffles

TUBE MATERIAL
Titanium

CORE/BAFFLE MATERIAL
Stainless Steel

LENGTH (INCHES)
5.1

WEIGHT (OUNCES)
6.56

TOOLS
Proprietary

SOUND
WEIGHT
ACCURACY
EASE OF MAINTENANCE
VERSATILITY
SIZE
PRICE

NOTES:

DESIGN: Skirted K-baffle with dedicated expansion and normalization chambers. Skirted baffles prevent lead and carbon from migrating between the stack and tube, allowing the stack to slip out easily for cleaning. Index tangs on each baffle ensure ports are aligned for best performance. Cerakoted tube, cap, and mount are nitrided steel. Solid construction and materials. Will last forever.

PERFORMANCE: Pistol; light FRP; tone is a "tick, tick." Weight is noticeable. Rifle; very quiet, but noticeable POI shift and middle-of-the-pack accuracy.

CLEANING: Damned proprietary wrench for endcap. Baffles come loose with a push when dirty. Uncaptured tube can twist off mount. Baffle arrangement is pretty obvious, but engraved numbers would be welcome.

GEMTECH GM-22

MSRP
$395

URL
www.gem-tech.com

CALIBERS
.22LR, .22WMR, .17HMR, .17HM2

CONSTRUCTION
Monocore

TUBE MATERIAL
7075 Aluminum

CORE/BAFFLE MATERIAL
Aluminum with Titanium Threads

LENGTH (INCHES)
5

WEIGHT (OUNCES)
2.47

TOOLS
1/4-inch socket driver

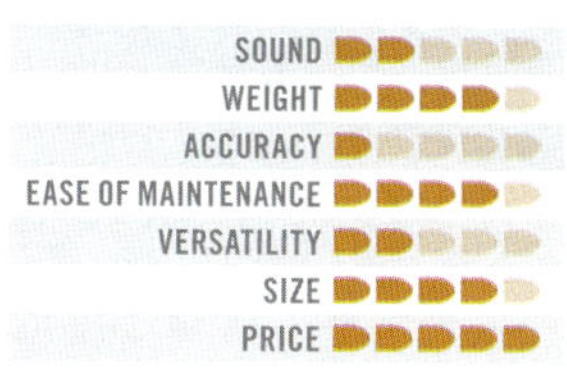

NOTES:

DESIGN: Aluminum monocore. Simple. Only two pieces: the core and the tube. Tube screws onto the core at the muzzle-end. Titanium muzzle threads are embedded in the core. Uncaptured tube can unscrew from core. Gemtech can upgrade an old Gemtech Outback can to the user serviceable GM-22 core for $150.

PERFORMANCE: Heavy FRP on pistol. Deep, bassy report. Handles well.

CLEANING: Threads tend to foul with carbon and lead since they're at the muzzle end. Takes significant force to break the core free after 500 rounds. We love that a ¼-inch socket head is the only tool needed to remove the core.

GEMTECH MIST-22

MSRP
$550

URL
www.gem-tech.com

CALIBERS
.22LR

CONSTRUCTION
Integral Monocore

TUBE MATERIAL
Aluminum

CORE/BAFFLE MATERIAL
Aluminum

LENGTH (INCHES)
16.1

WEIGHT (OUNCES)
18.6
(7.3 lighter than stock 10/22 barrel)

TOOLS
None

NOTES:

DESIGN: Aluminum monocore suppressor is permanently attached to a 9-inch steel 10/22 v-block rifle barrel and covered with an aluminum tube. Versatility score suffers since it only works with 10/22s. Outer tube diameter is same as a heavy barrel (0.920 inch); works in any heavy barrel 10/22 stock.

PERFORMANCE: "Tick-Tick" report. Combination of decent sound reduction, weight reduction and improved accuracy compared to stock barrel makes this one of our favorites. Bulk ammo will go supersonic, though. With subsonic ammo, it's night-before-Christmas quiet. Older version was ported to make normal velocity ammo subsonic.

CLEANING: Tube screws right off after 500 rounds. Threads would get fouled in older, ported version leading to seized-on tube. No issues with new, non-ported version. But gotta run subsonic ammo for best noise reduction.

GRIFFIN ARMAMENT CHECKMATE QD

MSRP
$449

URL
www.griffinarmament.com

CALIBERS
.22LR, .22WMR, .17HMR

CONSTRUCTION
Hybrid Monocore

TUBE MATERIAL
6061 Aluminum

CORE/BAFFLE MATERIAL
Aluminum

LENGTH (INCHES)
5.8

WEIGHT (OUNCES)
5.57

TOOLS
Proprietary

NOTES:

DESIGN: Aluminum monocore with three-lug QD muzzle device. Hybrid design has baffle-like blast face and chamber for reduced FRP. QD seems good for swapping between pistol and rifle. Added complexity and cost of QD isn't worth the trouble on rimfire, especially when extra adapters are $30 each.

PERFORMANCE: Initial groups strung out like Robert Downey Jr. in the '90s. Defective QD mount to blame. Two- to 3-foot POI shift depending on which of the three positions of the three-lug muzzle device the can is engaged in. Once the faulty mount was replaced, good accuracy and good sound reduction on rifle; the loudest can on the pistol. Zero FRP.

CLEANING: Proprietary tool needed to remove basecap. Endcap unscrews by hand. Included pusher tool required to remove core after 500 rounds.

EL CAMINO

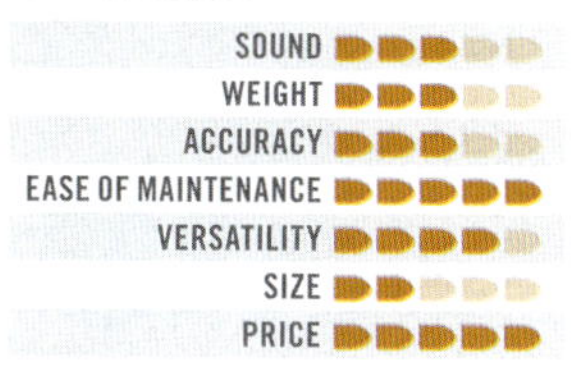

MSRP
$399

URL
www.liveqordie.com

CALIBERS
.22LR, .22WMR, .17HMR, 5.7x28 FN

CONSTRUCTION
M Baffles

TUBE MATERIAL
PVD- Coated Titanium

CORE/BAFFLE MATERIAL
Stainless Steel

LENGTH (INCHES)
5.9

WEIGHT (OUNCES)
4.48

TOOLS
½-inch socket

SOUND
WEIGHT
ACCURACY
EASE OF MAINTENANCE
VERSATILITY
SIZE
PRICE

NOTES:

DESIGN: All seven M-baffles are identical. Baffles form a tube-within-the-tube to keep crud from welding baffles to the tube. Steel muzzle threads are permanently installed in tube; no unintentional unthreading. Slim, long shape is barely bigger than .22/.45 barrel. Short thread section; includes spacer for barrels with longer muzzle threads. Average performance with excellent features and top-shelf materials at a below average price make this one of our favorites. Design bears a striking resemblance to the SIG SRD22X.

PERFORMANCE: Very light; handles great on pistol and rifle. Report is "hiss-tk." FRP returns quickly between mag changes. Sound levels and accuracy are middle of the pack.

CLEANING: Front cap flourish looks decorative, but accepts a common ½-inch socket for removal. No proprietary tools needed. Baffle stack fell out with 500 rounds of buildup. Suspect it could go well over 1,000 rounds between cleanings. No aluminum means the whole thing can go in an ultrasonic cleaner.

RUGER SILENT-SR

MSRP
$449

URL
www.ruger-firearms.com

CALIBERS
.22LR, .22WMR, .17HMR

CONSTRUCTION
Push Cone / Slant Baffles

TUBE MATERIAL
Titanium

CORE/BAFFLE MATERIAL
Stainless Steel

LENGTH (INCHES)
5.3

WEIGHT (OUNCES)
6.28

TOOLS
Proprietary

SOUND
WEIGHT
ACCURACY
EASE OF MAINTENANCE
VERSATILITY
SIZE
PRICE

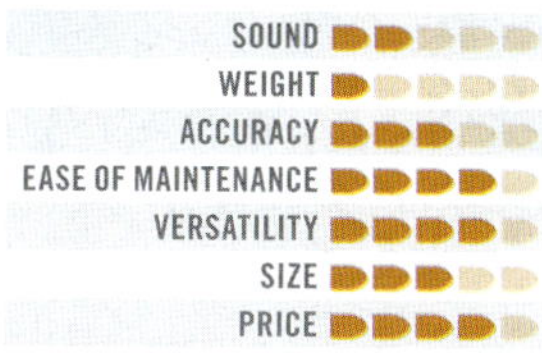

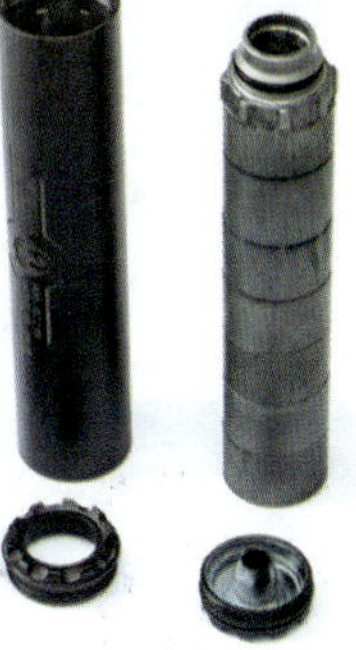

NOTES:

DESIGN: Uses Ruger-developed slant-nose, push-cone baffle system. Seals the interior and prevents crud from filling the space between baffle stack and tube. Extremely fine machining demonstrated in snap-together baffles. Muzzle mount locks in splines on the tube to prevent the tube from turning off the mount. One of the shortest, but on the heavier end of our test group.

PERFORMANCE: Almost no FRP, Hollywood "tick-sssss" tone. Good accuracy.

CLEANING: Proprietary tool for disassembly. Costs $20 to replace. Mount end has to be removed to pull core; threads get fouled and stick a bit, but baffle stack slips right out after wrestling cap off. Simple shapes of steel baffles easy to clean. We confused the blast baffle with a primary baffle at first. Almost used a hammer to mate them up before we noticed the difference.

SIG SAUER SRD22X

MSRP
$445

URL
www.sigsauer.com

CALIBERS
.22LR, .22WMR, .17HMR/.17 Mach II

CONSTRUCTION
M Baffles, Steel Mount and Endcap

TUBE MATERIAL
Titanium

CORE/BAFFLE MATERIAL
Stainless Steel

LENGTH (INCHES)
5.8

WEIGHT (OUNCES)
5.15

TOOLS
Proprietary

SOUND
WEIGHT
ACCURACY
EASE OF MAINTENANCE
VERSATILITY
SIZE
PRICE

NOTES:

DESIGN: Steel M-baffles and Titanium tube with steel muzzle threads. Muzzle threads are captured in the tube; twisting off the can won't leave the mount/endcap on the muzzle. Comes with 1/2-28 and M9x.75 rear mount to fit SIG Mosquitos. Design bears a striking resemblance to the Q El Camino.

PERFORMANCE: Excellent sound reduction, especially on a pistol with regular velocity ammo. Heavy FRP, though. Holds decent groups and cools down quickly between strings. Light; little effect on pistol handling.

CLEANING: Proprietary endcap tool. M-baffle setup works well and keeps crud from sticking to the tube walls.

SILENCERCO OSPREY MICRO

MSRP
$599

URL
www.silencerco.com

CALIBERS
.22LR, .22WMR, .17HMR, .17 WSM

CONSTRUCTION
Tubeless Square Baffles

TUBE MATERIAL
N/A

CORE/BAFFLE MATERIAL
Stainless Steel

LENGTH (INCHES)
4.6

WEIGHT (OUNCES)
6.77

TOOLS
5/32-inch allen

SOUND
WEIGHT
ACCURACY
EASE OF MAINTENANCE
VERSATILITY
SIZE
PRICE

NOTES:

DESIGN: Tubeless, square baffles stack together for two lengths (4.6 and 3.1 inches). Changing length is easy. Offset design doesn't block pistol sights, but none of the round tube cans did, either. Steel baffles are tough; survived a baffle strike without a scratch after accidently cross-threading on QD adapter. QD setup works well, but initial install is a PIA as the QD adapter needs to be timed with shims for each gun.

PERFORMANCE: All sound data gathered in long configuration. Bassy report. Some FRP on pistol. Feels heavy in full-length configuration on pistol. Short configuration is loud, but still hearing safe.

CLEANING: Once the QD mount is installed, disassembly and length swaps only require an Allen wrench. No tube to deal with. Stack falls apart once head screw is removed no matter how many rounds through it. Big, open baffles are super easy to clean. Only goes back together one way. We'd like it more without the QD setup.

SUREFIRE RYDER 22-S

MSRP
$469

URL
www.surefire.com

CALIBERS
.22LR, .22WMR, .17HMR

CONSTRUCTION
Pignose Baffles

TUBE MATERIAL
Aluminum

CORE/BAFFLE MATERIAL
Stainless Steel

LENGTH (INCHES)
5.4

WEIGHT (OUNCES)
5.08

TOOLS
Proprietary

SOUND
WEIGHT
ACCURACY
EASE OF MAINTENANCE
VERSATILITY
SIZE
PRICE

NOTES:

DESIGN: Novel, steel pig-nose baffles nest together to keep lead/carbon particles from caking on the tube. Patented muzzle section is splined to index with tube; no twisting the tube off instead of the whole can. Comes with a spacer for longer threaded muzzles and an insert to help push the stack out.

PERFORMANCE: "Thwap" tone. Noticeable FRP. Quieter on rifle versus pistol.

CLEANING: Proprietary spanner needed for disassembly. Baffle stack doesn't fall free after 500 rounds. Took a little pushing. Steel baffles are GTG in an ultrasonic cleaner; intricate shape is tedious to clean by hand. Numbered and rotation indexed baffles are easy to restack.

THUNDER BEAST ARMS 22 TAKE DOWN

MSRP
$395

URL
thunderbeastarms.com

CALIBERS
.22LR, .22WMR, .17HMR, 5.7x28mm

CONSTRUCTION
Conical Baffles

TUBE MATERIAL
Titanium

CORE/BAFFLE MATERIAL
Stainless Steel

LENGTH (INCHES)
5.6

WEIGHT (OUNCES)
5.71

TOOLS
15/16-inch socket (included)

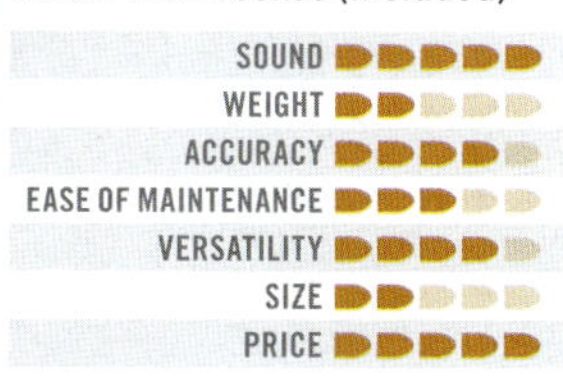

NOTES:

DESIGN: Snap-together, rotation indexed conical baffles provide above average sound reduction and accuracy. Lack of anti-rotation tube is a drag. But once the muzzle end gets crap in the threads, it actually helps the endcap from unscrewing from the tube. Overall sound, accuracy, features, and ease of maintenance make this our top choice regardless of price.

PERFORMANCE: Quiet "thwap" tone. Longer and heavier than average, but overall, the quietest when averaging rifle and pistol sound measurements. Above average accuracy.

CLEANING: Comes apart with included 9/16-inch wrench. Adjustable wrench works in a pinch. Stack slides out with a push after 500 rounds. All steel and Ti; chuck it all in the ultrasonic cleaner. Baffle stack easily assembled and timed with tabs and index marks on baffles, then slipped back in tube.

ACTUAL INTEGRAL

We Dig Into the Suppressed 5.56 Integra from Gemtech

By Dave Merrill

In mid-April 2016, USSOCOM officially started solicitation for a Suppressed Upper Receiver Group (SURG). The requirements for the contract were extremely lofty and eventually the program was scuttled. Then reborn. Then got mucky again. Thus is the nature of government procurement. Several manufacturers submitted samples, but at the time of this writing, Gemtech is the only one who brought their offering to the commercial market — almost.

There are a couple of differences between the SURG upper and the Gemtech Integra, but functionally it boils down to this: The Gemtech SURG upper has just over an inch more rifled barrel, and the overall length of the upper was reduced by 1.7 inches due to government requirements.

To be sure, an integral silencer isn't a normal first NFA purchase. When starting out, buyers typically want something with more perceived utility, such as one of the myriad of multi-caliber silencers. Later, they branch out into specific calibers and guns. An integral is likely at least four or five slots down the list, if one is ever purchased at all.

SINGLE-STAMPING ACROSS STATE LINES

As an individual, before crossing state lines with most types of legally owned federally regulated firearms like machineguns and short-barreled rifles (SBR), you need an approved BATFE Form 5320.20. Though it's very dependent on your examiner, it's not unheard of for an approved form to take four to six weeks to arrive. We know numerous people who were unable to bring their goodies with them to a class or other event for lack of approval. For reasons that we don't understand and don't want to ask about, this annoying law about interstate travel applies to all federally regulated firearms, sans two categories: Any Other Weapon (AOW) and silencers.

Provided that silencers are legal in your destination state, you can cross state lines with no notice and no permission. Hence, the advantage of the one-stamp silenced rifle.

While a one-stamp rifle has certain legal benefits in the United States, there are some additional perks too. Namely, a manufacturer can build it from the ground up with complete integration in mind. The Gemtech Integra certainly isn't the first gun with a built-in suppressor. In recent years we've seen several new models come to the market, such as the SilencerCo Maxim-9.

Many integral rifles have a silencer monocore machined into or permanently attached to the barrel. Some psuedo-integrals simply have

The titanium monocore is easily removed with a 3/8- inch socket wrench.

The four angled ports in the end of the monocore make the Integra self-tightening.

a silencer welded to a shorter barrel. Both of these methods are easy ways to reach that 16-inch minimum of a non-federally regulated rifle. Gemtech took a different approach with the Integra, opting to weld a sleeve to the barrel that encompasses a removable titanium monocore. With the twist and turn of a ⅜-inch socket wrench, the core comes out for cleaning and servicing. The core itself features a sharp edge that cuts any accumulated carbon during removal. There are four angled ports machined into the end of the monocore, making it self-tightening.

Get an endcap strike with a permanently attached core and you're SOL, whereas the core on the Integra can be readily replaced. Additionally, have you ever try to run military bore patches through a silencer? We don't recommend it.

The ability to remove the entire monocore assembly leaves room for upgrades by Gemtech into the foreseeable future. Perhaps down the line, we'll see silencer upgrades for this rifle.

We admit, seeing that this was a pistol-length gas system with a monocore gave us some pause. But we withheld judgment until we got it to the range.

5.56 silencers are loud. The round's high pressure and velocity just aren't conducive to that movie-quiet experience that one can attain with subsonic .300blk or .22LR. Though in theory subsonic 5.56 could be used to further muffle the report, its terminal ballistics leaves a lot to be desired. If you ever want to see a face worthy of being featured in a sad Sarah McLachlan ASPCA commercial, we suggest observing someone who just shot their brand-new 5.56 silencer for the first time — and who has never experienced one firsthand. You can practically hear "Angel" playing in the background as the tears of sadness and regret roll down their faces.

Aside from the caliber being loud in general, the suppressed AR-15 can be unusually troublesome at times. Over the last six decades of development, we've seen a series of engineering challenges updating and upgrading the AR-15 to fit current mission requirements. These unplanned impositions ultimately made the AR the modular monster that we currently know and love. Dealing with silencers was but one of those ongoing obstacles.

By containing the expansion of hot gases at the muzzle in order to bring down the sound, a silencer can significantly increase backpressure into the system. This can mean malfunctions from excessive bolt speed, bolts attempting to unlock far too early, more pronounced expulsion of gas out of the ejection port, and higher rates of fouling, not to mention a face full of burnt powder. After these last several paragraphs, one may get the impression that we hate suppressed ARs, but nothing could be further from the truth.

These problems all have solutions in the form of all manner of specialized adjustable parts, weights, and differing silencer designs. Modular monster indeed. The Integra does something different enough that they patented it, and now that said patent has gone through we can share it with you. It's gone largely unnoticed, mostly because it's welded to the barrel and hidden by the monocore sleeve. The Integra has a bore evacuator. Like the main gun on a tank. Here's how it works:

› The projectile first passes the primary gas port, and some of the gases travel down the gas tube to begin the cycling of the rifle as usual.

› Shortly after this, the bullet passes two additional gas ports. These ports lead into a chamber that surrounds the barrel expressly for this purpose, and a fixed amount of high-pressure gas is held.

› Once the projectile leaves the barrel, the resulting vacuum sucks the gas in this chamber out toward the muzzle.

Effectively, the bore evacuator delays the movement of the gas, giving it slightly more time before it follows the path of least resistance. The end result is that the majority of the extra gas goes through the muzzle end instead of back-flowing through the barrel and the effects that come with it. When we asked Gemtech if the bore evacuator reduces the velocity of the ammunition shot, similar to a ported barrel, they told us that if anything it actually increases the FPS. Since the bore evacuator just delays the gas rather than venting it out like an open port in a barrel, this makes a good deal of sense. In order to test it for ourselves we would need another Integra sans bore evacuator, since there are numerous factors that can affect muzzle velocity.

BUILD QUIRKS

But not everything is all daisies. Because the gas block and sleeve are welded into place, you won't be swapping out your handguard system anytime soon. Thankfully, instead of a lower-end handguard they went with a Seekins MCSR V2 handguard. If you somehow ended up with a KeyMod and wanted an M-LOK or vice versa, anything in the Seekins MCSR line can be swapped out. Overall, the modular nature of the AR-15 becomes a skosh less modular with the Integra.

The Dead Foot Arms folding stock makes for a small package while maintaining functionality.

Take note that depending on who makes your M-LOK accessory attachment, you may have to hit the hardware store for some slightly smaller screws to avoid marring the silencer shroud. While everything fits under the handguard, you don't have an awful lot of room on the front half. We had to shave down the screws on the Magpul MVG, and at the end of the day it was easier to remove the rail to add grip covers a la the Troy TRX rail.

Since the upper receiver itself promises to be perennial on the Integra, Gemtech thought it would be best to go with a high-quality known quantity, the VItor MUR-1A. It's the same internally as any other

Gemtech's patented bore evacuator

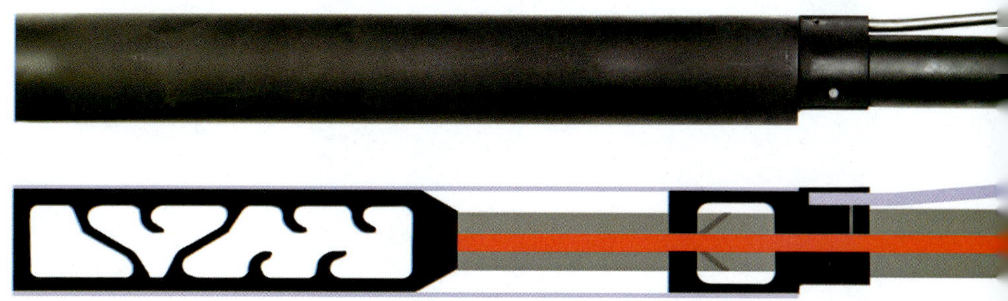

As shipped, all it needed was a set of chrome truck nuts to complete the package. Yeah, no.

Mil-spec upper, but rocks thicker walls, is fully heat treated, stress relieved, and even subjected to the mysticism of cryogenic freezing.

THE JERSEY SHORE

Though normally the Integra is sold as a separate upper, our particular example has a corresponding Vltor VRA-R lower receiver, complete with Gemtech manufacturing marks. In addition to the recessed extra large magazine release, the VRA has a set of ambidextrous limited-travel QD sling swivels inset into the lower itself. A lightweight two-stage flat trigger from CMC Triggers completes the package.

The Gemtech Integra we received was gaudy to say the least. Heavy with highly contrasted branding, chrome, and polish, this is damn near the rifle equivalent of a bro with a Tap-Out T-shirt and a fistful of energy drink. We wouldn't have been surprised to see it snorting lines with people who could be cast on *The Jersey Shore*. Start your fist pumping now.

They say beauty is just a light switch away or, in our case, a can of Krylon and a box of replacement parts. Like refacing a lime green 1970s kitchen, a change of color can make a world of difference. We removed and replaced most of the chrome, and swapped out the more egregious branding. Same rifle in a different dress, for sure. But 1,000 times more palatable. Like wearing a Rolex on a NATO strap, there's just something special and fun about using $12 in paint on an expensive rifle.

We won't apologize for focusing on some aesthetics, as rarely does someone cross a street to meet your personality. Since the Dead Foot Arms system adds several inches to the length of pull if used with a standard stock, a shortened receiver extension and buttstock from an LMT PDW stock tightened up that distance. We also liked how the Dead Foot Arms can be fired when the stock is folded, however unlikely it is that particular feature will be needed. In order to achieve this feat, the crew at Dead Foot Arms uses a truncated bolt carrier and a dual spring setup.

AT THE RANGE

It's always nice to know that you're the first one to shoot a rifle. And we know we were because the CMC trigger wasn't resetting. Every other trigger we tried worked

fine, while the CMC wouldn't reset with a regular upper either. So, we dropped in an ELF trigger instead. Once you start playing mucky muck with specialized buffer spring assemblies like you have to with PDW stocks, you can run into issues (for a full breakdown, see our PDW Stock Buyer's Guide in Issue 30). Locking back on an empty magazine was inconsistent before a little spring tuning was performed. Neither were issues related to Gemtech — and you can put the Integra on any 5.56 lower that you want.

Gas to the face? None. Zero. Nadie. Nada. The bore evacuator and tuned gas system are great at their job. On a similar note, usually our internals and magazines get scummed up quickly when shooting suppressed due to the dirty backpressure. We are happy to report that nothing seemed any dirtier than a regular-ass unsuppressed 16-inch AR.

For science, we also tried the Integra without the monocore installed. This isn't recommended because, well, it doesn't work. This rifle is specifically tuned to run with the pressure that the silencer produces. No monocore, no worky.

We tested accuracy with match Prvi Partizan 75-grain ammunition, and the Integra printed a five-shot group at just under 1.5 inches at 100 yards.

Regarding sound? We'll rely on folks with $40,000 rigs to make that determination, but Gemtech advertises 131 decibels. This is about on par with an unsuppressed .22LR rifle, and that seems to hold true. We can say that the ejection port noise is greatly reduced, and the Integra was quieter and produced less flash than our other 10.3-inch guns with several different silencers. A nice surprise, since a monocore design isn't widely regarded as being the most effective.

LOOSE ROUNDS

We're not completely sure about the first integral, but arguably the most famous is the Heckler & Koch MP5SD. And the coolest? The OSS' Nazi-snuffing High Standard HDM. As to where the Integra falls onto the list of viable guns — it has a lot going for it. We wish there was more choice regarding handguards. The enclosed front end largely negates the ability to further refine the gas flow, reducing your options to an adjustable bolt carrier gas key like those sold by Rubber City Armory. With that said there's still some flexibility on the rear end of the equation with buffer and spring weights.

When we started, we looked at the Gemtech Integra like a short-barreled rifle with a silencer attached, but it's more than that. The thoughtfulness of design, the fine-tuning, and how it sounds all add up to a package more than just a sum of its parts. Actual. Integral. Hot damn.

Grouping just under 1.5 inches at 100 yards, accuracy is acceptable but not exceptional.

A relatively short, quiet, soft-shooting suppressed rifle you can cross state lines with without asking permission from Uncle Sugar? We'll be in our bunk. **R**

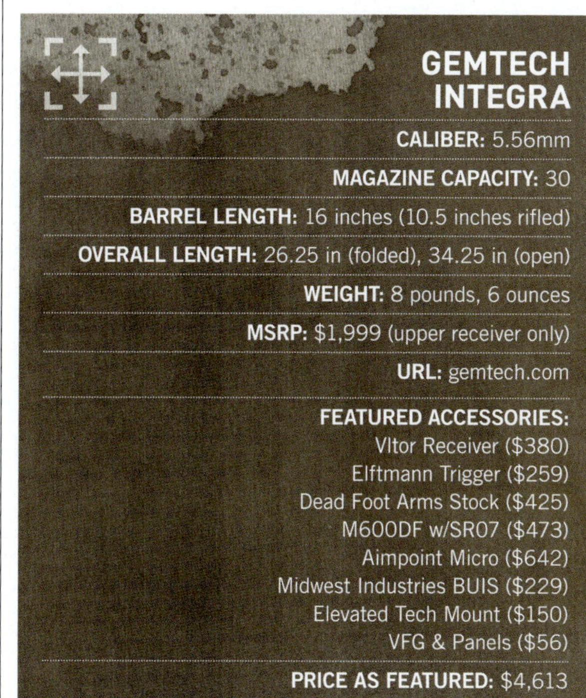

GEMTECH INTEGRA

CALIBER: 5.56mm

MAGAZINE CAPACITY: 30

BARREL LENGTH: 16 inches (10.5 inches rifled)

OVERALL LENGTH: 26.25 in (folded), 34.25 in (open)

WEIGHT: 8 pounds, 6 ounces

MSRP: $1,999 (upper receiver only)

URL: gemtech.com

FEATURED ACCESSORIES:
Vltor Receiver ($380)
Elftmann Trigger ($259)
Dead Foot Arms Stock ($425)
M600DF w/SR07 ($473)
Aimpoint Micro ($642)
Midwest Industries BUIS ($229)
Elevated Tech Mount ($150)
VFG & Panels ($56)

PRICE AS FEATURED: $4,613

GEMTECH

Back in the dark ages of the 1990s, when Will Smith was still a young comedic actor, Gemtech began working on a special-purpose suppressor. It had to be as small as possible, lightweight, and quiet. A lot of special and one-off items have been made for downed pilot survival kits — the Aurora would become one of them. Getting your grubbies on one of the originals was hard, but now there's a newer and better one out there: The Gemtech Aurora II.

The first suppressor to be made in-house by new Gemtech parent company Smith & Wesson, these Auroras will actually hit the larger open market.

AURORA II

SIZE MATTERS — JUST NOT THE WAY YOU THINK

By Dave Merrill

BETTER, FASTER, STRONGER

Several changes and improvements went into the Aurora II. First and most obvious is the exterior sleeve. It's slotted and then slightly bent in at the middle. This makes gripping the silencer to take it on or off exceedingly easy. Secondly, it's tandem threaded and bidirectional. Have a 9mm pistol threaded M13.5x1LH? There's an end for that. Have a 9mm pistol in the preferred thread pattern of the only nation to put a man on the moon, ½x28? There's an end for that too.

Each end comes complete with internal thread protectors and wrench flats that conveniently fit into an AR armorer's tool.

The internals are a palindrome of wipes and spacers, so it makes zero difference which end of the Aurora II you use.

Less obvious are some of the internal changes of the Aurora II versus the original. Firstly, it uses wipe material that is 85A durometer polyurethane, much tougher than the original 65A material. Secondly, the spacers are now bidirectional. While there were several different versions of the original Aurora, depending on what year it was produced, many had spacers with different openings on either end. The idea was that having the muzzle-end hole slightly larger gave more space for a wipe to cave into as it wore.

After a lot of testing and trial and error, engineers at Gemtech determined that simply having the larger hole on either side actually increased wipe longevity. The change in material and spacer design made for a tougher suppressor. How much? The original Aurora had an anticipated wipe life of 10 rounds when mounted on a service weapon — and the Aurora II: 40.

Did we mention that the Aurora II is small and light? At just 3.3 inches long and a scant over 3 ounces, this can won't crack any scales. Because it's so small and light, no booster or Nielsen device is needed even when used with a Browning tilt-action barrel.

We don't care what consenting adults do, but this kind of docking is absolutely not recommended.

OLD TECHNOLOGY FOR NOVEL USES

Normally when we think of silencers that use wipes, old Mitch Werbell designs come to mind. Wipes can be made from all manner of materials, so long as it's pliable enough to allow a projectile to pass through but seals up fast enough to trap the explosive gases behind. Those self-healing targets you throw downrange work much in the same manner.

There are definitely downsides to wipes; they can't be used very much before replacement, accuracy can get thrown off considerably because the bullet passes through the material, and generally they aren't as well suited for modern weapons when compared to present-day options.

But of course there's a but. You can make a silencer exceptionally small with wipes and grease. As mentioned wipes help keep the gas sealed in, and the grease is an ablative media — in other words, a temporary performance enhancer. If you've ever heard about a suppressor being run "wet," that's in reference to an ablative media being used.

After the Aurora II is packed with grease and fresh wipes, it can be tucked away for years without any worries of leakage or evaporation.

THE PROJECTILE PREDICAMENT

While you may think that the Aurora II might make for a nice little nightstand or carry can, you'll have to pump the brakes here. Because the projectile passes through several wipes and some grease along the way, even with the X-aperture in the wipes themselves, the bullet strikes material on its way down range. This means hollow points or any other expanding ammunition are a no-go. To quote Gemtech, "guaranteed endcap strike."

Slightly further complicating matters is that only one side of the silencer threads off — if both sides came off, an endcap strike could be a quick fix.

While our service men and women the world over largely carry ball ammunition in their pistols,

The internals are deceptively simple — and this is the cleanest they'll ever be.

we won't pretend this is the best possible choice for defensive application.

AT THE RANGE

Exactly how long the wipes will last in the Aurora II goes far beyond the user manual. Supersonic rounds through a short pistol will wear out wipes faster than subsonic through a duty-sized gun. The more powder you have burning and blasting from the muzzle, the shorter the life span.

The original Gemtech Aurora was designed around subsonic Israeli 158-grain ball ammunition. We can confirm this load works very well, as does Freedom Munition 165-grain HUSH. Ammunition with truncated cone projectiles, like a lot of 147-grain on the market, will reduce the life span of the wipes.

Because we wanted to play with extremes, we used supersonic 124-grain NATO ammunition in a tricked out micro-murder Glock 43, and 165-grain HUSH ammunition out of a Glock 17. Differences were night and day. With a 165-grain from a full-size pistol, we were able

The guts come out charred and smoked to perfection.

to unload a full box of 50 before the wipes gave way — from the G43, about 17 rounds.

Even when the centers of the wipes were blown out, there was still some (albeit small) suppressive effect.

While subsonic ammunition can make any silencer sound pretty good, what surprised us about the Aurora II is that it worked exceptionally well with supersonic ammunition too. Once the silencer gets heated up and the grease starts burning off, it can get quite smoky at the range — so definitely ensure you have good ventilation if you're partaking indoors.

When the wipes go, the centers are completely removed. If you take off your Aurora and can see through it, it's time for new guts.

Accuracy was very dependent on the pistol and load as well; remember we're talking about blowing though a total of an inch of polyurethane after exiting the muzzle. Supersonic ammunition tended to deflect less, but inconsistent wear of the wipes may also have some effect.

LOOSE ROUNDS

The original Aurora may have been developed for pilot escape and evasion kits, but it soon found roles with other parts of the defense apparatus. It isn't terribly hard to imagine one of these being employed against a sentry's skull, making their brain the final baffle.

Enough of these went to specific SOCOM units that Gemtech was tasked with providing training on field-expedient improvised insides. They tried everything from cut up Glock-branded stress balls to sheepskin to sleeping pads. Ultimately the original polyurethane was the best, but leather ended up being something of a viable alternative — no huge surprise as many early silencers sported leather wipes.

Compared to full-size pistol suppressors, the Aurora II is maintenance heavy and has a whole lot of ammunition restrictions. But for the original intended role and other government special uses that came later down the line? Unparalleled. The very best part of the Aurora II isn't simply the size or how quiet it is, but the fact that you can actually own and use one yourself without crashing an F/A-18. **R**

GEMTECH AURORA II

CALIBER:
9mm

MATERIAL:
Aluminum and stainless steel

LENGTH:
3.3 inches

DIAMETER:
1.125 inches

WEIGHT:
3.2 ounces

MSRP:
$399

URL:
gemtech.com

MAKING REPLACEMENT WIPES

The best suggested practice for replacing the wipes in your Aurora II is to send it in to Gemtech. For $30, they'll put in brand-new factory wipes, grease it up properly, and send it on back. However, if you're a type 02/07 Special Occupational Taxpayer (SOT) FFL, you can legally make new wipes without any questions or hassle. Legal stuff aside and without further ado:

TOOLS & MATERIALS

- 15⁄16-inch hammer driven arch punch (McMaster-Carr #3427A23)
- Craft knife
- Petroleum jelly (premium synthetic grease is a viable, yet unpreferred, alternative)
- BFH
- Gloves
- 12x12-inch sheet of 1⁄8-inch-thick polyurethane (PU) rubber, 80A hardness (McMaster Carr #8789K21); enough for well over 100 wipes, but you can buy in any quantity

Notes: OEM Gemtech wipes are water-jet cut. However, these instructions closely follow Gemtech-founder Dr. Phil Dater's instructions for original Aurora wipes. You also have some options regarding tools and materials. For example, the PU hardness for OEM wipes is 85A, which we found more difficult to source. The 80A is an easy alternative with only slightly reduced longevity.

Additionally, if you already have a 1-inch arch punch (McMaster-Carr # 3427A24) that'll also work — but you'll need to use a dowel to press fit. Both sizes are equally as effective. Another optional tool is a sharp ½-inch cold chisel (McMaster-Carr #3575A22) in lieu of a craft knife; while it isn't more effective, it does look nicer. Neatness doesn't count terribly much here in regards to performance though.

1. Disassemble the Aurora II. Destroy and dispose of spent wipes. Clean spacers if needed (being a little dirty is no problem).

2. Using your BFH and arch punch, cut eight wipes from your PU. A cutting board makes this job easier and will also keep your punch sharp for a longer period of time.

3. Use your craft knife (or ½-inch cold chisel) to cut an "X" through the center of the wipes. Pair each wipe with another, using a dab of petroleum jelly to stick them together. Each cut "X" should be offset 45 degrees from one another.

4. Pack the outside and inside channels of your spacers with the petroleum jelly. Also apply jelly to the top and bottom of each spacer for the best seal with the wipes.

5. Assemble the silencer. The guts consist of eight wipes stacked together in four pairs, with three spacers. The order of assembly is wipes-spacer-wipes-spacer-wipes-spacer-wipes.

6. If you've used a 1-inch arch punch rather than a 15/16th, use a dowel to press fit. Replace the front cap and enjoy!

WARNING! The BATFE has notified manufacturers that suppressors and suppressor parts, including suppressor wipes, are federally regulated devices and can only be manufactured by specially licensed firearms manufacturers.

1

2

3

4
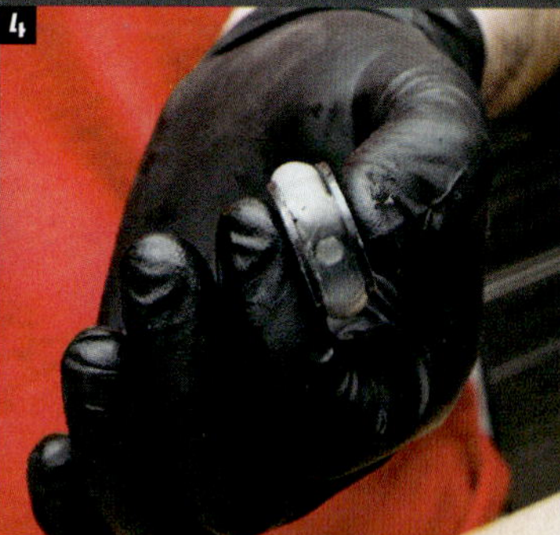

5
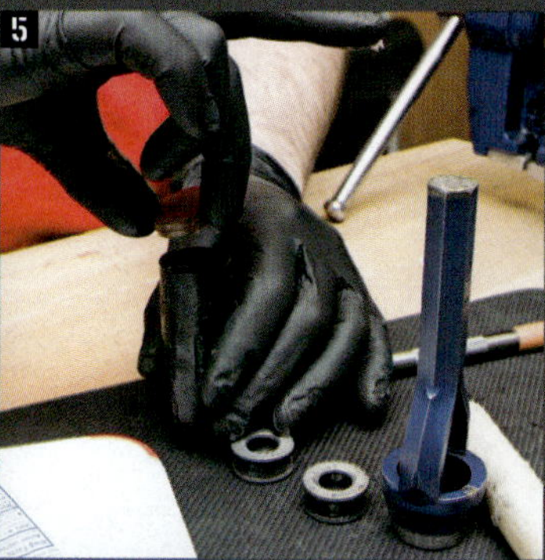

6

GETTING KURZ

We Let You Know What They're Good For

By Dave Merrill

WITH IT

If having big guns indicates a tiny tool, we must all have bulging britches — because smaller everything is cooler when it comes to guns and silencers. There are fashions and trends with suppressors just like there are with guns, cars, jeans, and narcotics. Numerous criteria factor into selecting a silencer, such as size, weight, durability, warranty, flexibility, and, you know, how much noise is actually produced — exactly which aspect is the most important to the public at large seems to change every few years.

Ten years ago, it was durability. It didn't matter if a silencer weighed a ton, as long as it was tough as sh*t and [kinda] quiet. Titanium cans weren't as popular, because despite the lower weight, they weren't as durable — when titanium heats up it erodes and flakes off.

Fast-forward a couple years and weight became the most important factor. Titanium was *in*.

A few years later, the multi-caliber silencer was the most popular. The one-size-fits-most that worked "pretty OK" at everything but wasn't exceptional at anything. As a first silencer, some of them still remain decent choices.

More recently the modular silencer has taken hold. Want it longer and quieter? Keep it long. Shorter and louder? Just configure it that way.

And right now, in early 2018? The subject of this article: K-cans.

STUBBY. SHORTY. NUBBY. KURZ.

Kurz is German for short; many devices have shared this moniker over the years, both by their actual product names and via slang. Modern "full-size" rifle-caliber suppressors range from 7 to 10 inches, fairly small compared to their 1980s equivalents. We're still not sure whether John Arthur Ciener should be proud or ashamed. For the purposes of this article, we arbitrarily used 6 inches as the cutoff for a Kurz (K) rifle silencer.

While size and weight of full-size rifle silencers vary, you add significant length to your rifle with any of them. Even those that explicitly state "only adds X inches beyond the muzzle" invariably require the use of an extra-long muzzle attachment. This isn't lying — not exactly — but it is disingenuous. Of course and as usual, we blame marketing teams.

Not only are you adding length, effectively turning your handy little carbine into a musket-length affair like your great-great-great grandfather may have used, but you're also significantly throwing off the balance. It seems everyone loves a silencer on a 16-inch barrel — until they use one on a short barrel rifle (SBR). Sure, you've shaved off a few inches, but compact you aren't.

One way you can still get some of the benefits of a silencer without throwing a ton of weight or length on the end of a gun is with a K-can. But they aren't for everyone. To be clear, this isn't a buyer's guide: this article will help you understand what to expect with one, and gives you the tools to weigh the pros and cons for yourself.

On the face of it, the concept of a K-can is silly to some. After all, isn't the point of a silencer to be as silent as possible. While these stubby silencers do indeed cut down the noise, they're nowhere near as effective as their full-size brethren — double-so when paired with a short-barreled rifle. Triple-so if 5.56mm is involved. They look rad though, and that's justification enough for some.

A short silencer helps a stubby rifle stay small.

Many quality K-cans are hearing safe on carbines.

especially in a world of battle-worn finish and blinged-out blasters. And there are actually some reasons that you might want to nab yourself a K-can.

There are two basic ways to look at a kurz can:

- As an expensive linear compensator on a short barrel
- As a silencer you'll actually enjoy using on a full-size (16-inch+) rifle.

ROUNDING THE CORNERS

It shouldn't be a shock to anyone that short barrels produce a lot of flash and bang in rifle calibers. Out in the open, an SBR with a decent muzzle device may not rattle your teeth, but while indoors, inside a car, or with a team is another story altogether. And Lord help you if you're using a brake.

Linear compensators such as the Noveske KX-series or Troy Claymore don't reduce noise or blast, instead projecting it forward away from the shooter or anyone beside them. A stubby K-can will do the same, while reducing the noise signature to boot.

The original short-barreled AR silencer, the XM moderator built for the Colt Model 610/XM177, is only considered a silencer because of the absurdity of American law. While a proper XM moderator will decrease decibels almost to that of a 20-inch rifle (thus earning its legal status as a silencer), it mainly serves to cut down the blast. Furthermore, the XM moderator increased backpressure to ensure more reliable operation on these early shorties.

Don't expect miracles with 5.56mm, lest you set yourself up for disappointment, though a kurz does provide benefit in this role without tipping the scales too hard on the front end. It's for these reasons that K-cans are increasingly popular for military and law enforcement roles: They avoid blinding or deafening your teammates, they reduce the noise inside a structure allowing for better communication, and rifles can usually fit into patrol vehicles, gear lockers, and weapon racks with the silencer attached.

MAKE SUPPRESSED 16-INCH GUNS GREAT AGAIN

A 16-inch rifle with an 8-inch silencer easily makes a rifle longer than your great-uncle's Vietnam M14, in a harder to wield package. We're not even upset about that; it's impressive.

As suppressor technology marches on we can get the same performance from a large antique silencer in a smaller package. With that said, physics being what they are, a similar design with less internal volume will be louder.

Everything is a balance between weight, mobility, and capability. A silencer may not have to be completely silent — just quiet enough for your needs. You can give up a little noise for a shorter package. This is where we'd normally insert a

Fullsize can on a shorty and a Kurz can on a carbine are around the same size. Any silencer on a 20-incher makes it a full-fledged musket.

A little brother is just a squat version of its older sibling.

joke. We've found the current crop of kurz cans to be handier to use on 16-inch+ rifles than the heavy bastards we carried a decade ago.

We're not going out on a limb when we say that the 16-inch barreled rifle is by far the most popular size sold in the United States. The same benefits that an entry team attains from using a stubby silencer on an SBR directly translates to use with a standard carbine; better in every way sans those few extra inches of barrel. Reduced flash, decent balance in size and performance, and more likely to actually be hearing safe. Like Goldilocks gobbling all the porridge, you may find this combination just right.

FAT BASTARDS AND LITTLE BROTHERS

In the admittedly broad taxonomy of short silencers, there are a couple of base designs we can consider: fat bastards and little brothers.

Usually designed not to be regularly removed, a fat bastard makes up for its lack of length with girth. They're more likely to be direct thread (DT) silencers, but not always. While QD devices allow easy removal for transport or swapping, direct-thread silencers do have some advantages, especially when we're talking about the shorties. While some designs are better and more efficient than others, in general since DT designs have no need for special locking mechanisms or internal space to accommodate muzzle devices, they can be more efficient. However, this comes at a cost.

While many new DT silencers feature wrench flats and holes for blind pinning, generally we avoid permanently attaching a silencer unless we absolutely have to. Even though a DT silencer can be removed, functionally you should consider them long-term residents. Needless swapping is just bad for everyone.

Above: Your ammunition matters just as much as your muzzle device, silencer or otherwise, when it comes to visible flash signature

Top: Integral flash hiders can considerably cut the flash, especially from a short barrel.

Examples include the Delta P Design Brevis II and Innovative Arms Grunt-M. If you're the type who wants to fit a silencer under a handguard, these definitely aren't for you.

Like a miniature copy of their older and larger siblings, but shorter, little brother designs share aesthetics and features like quick-disconnect mounts you may already have. Examples include the Dead Air Sandman-K and the SureFire SOCOM556-Mini2. If there's already a Sandman-L or SureFire RC2 in your safe and you want to try some presto-change-o you should consider one of these.

MUZZLE BLAST AND MISCONCEPTIONS

Some say silencers are the best flash hiders out there. Like so many other oft-repeated axioms, this isn't true in every case. A silencer might be a decent flash hider, but frankly it's pretty damn hard to beat something like a B.E. Meyers M249F at reducing visible signature.

Especially when we're talking short cans on short guns, a $7 1960s flash hider holds its own remarkably well.

Hiding flash is about more than the muzzle device; ammunition selection and barrel length is just as important. Hotter ammunition with faster-burning powders often generate less flash, simply because the bulk of the powder burns inside the barrel before it has the opportunity to be expelled and ignited beyond the muzzle.

Whenever we demonstrate the flash signature of any weapon or attachment, we try to use the most consistently flashy ammo we can get our paws on. Federal American Eagle routinely produces a f*ckoff amount of flash from a bare muzzle — perfect for this role. Just have a gander at the image comparing it to IMI M193 with a 16-inch naked gun.

Bear in mind that the first shot from a silencer usually produces more flash than subsequent ones. This is due to the atmospheric oxygen inside the silencer burning off. As to how long you have before that next big flash? That largely depends on the silencer used. Some companies have integrated small flash hiders into their endcap designs to further reduce flash. Flash from follow-up shots is usually significantly less prominent.

LOOSE ROUNDS

If silencers are decriminalized via legislation like the Hearing Protection Act, you'd have no excuse not to have a handful of them. What makes a K-can so great isn't that they're amazing in any particular way aside from being small, but that they're OK at everything while being small. Just like the multi-caliber can, the strength of a stubby silencer is in its versatility. In the long list of NFA items, a Kurz rifle silencer ranks alongside an integral — it shouldn't be your first or second silencer, but it may make its way to the fourth or fifth place on the ledger. R

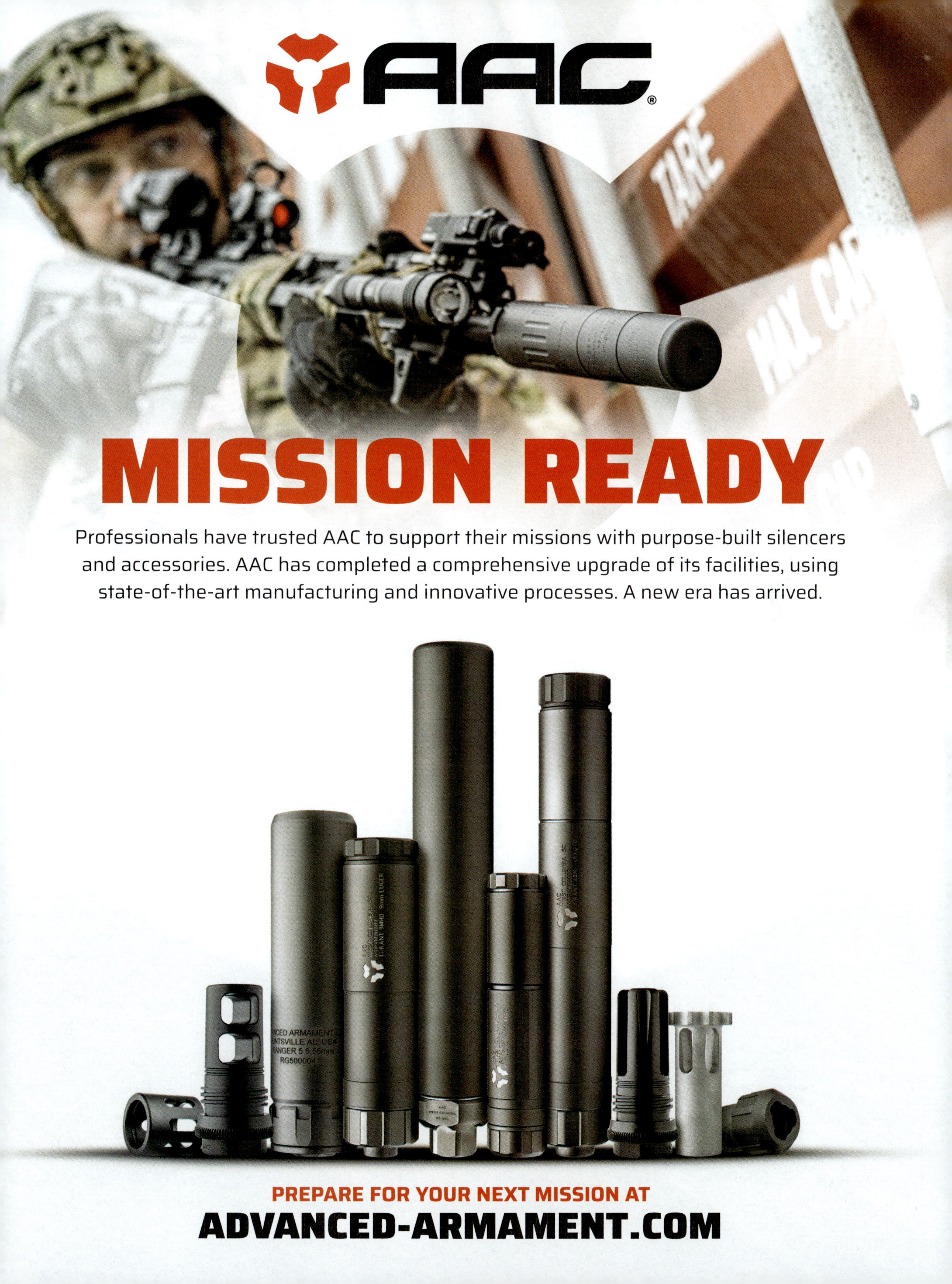
AAC®
MISSION READY
Professionals have trusted AAC to support their missions with purpose-built silencers and accessories. AAC has completed a comprehensive upgrade of its facilities, using state-of-the-art manufacturing and innovative processes. A new era has arrived.
PREPARE FOR YOUR NEXT MISSION AT
ADVANCED-ARMAMENT.COM

TAMING THE BEAST

WARNING!
When handling firearms, always observe safety rules and the precautions set forth in the firearm's owner's guide. Be certain that your firearm is unloaded and made safe before proceeding with this DIY.

Using Your Mary-Sue Suppressor on an AK Variant

By Dave Merrill

Want to silence an AK? You have an uphill battle. Back in Issue 26, we covered a pair of dedicated AK silencers — one from Gemtech and another from Dead Air. Since that time, a handful of other purpose-built Kalashnikov suppressors have been released, but what if you don't want to buy a silencer specifically for an AK? That's what this article is for.

Chances are, if someone only owns one silencer, it's either for 22LR or for 7.62N — this piece is for those who fall into the latter category. A 7.62N silencer as a "do all" is a very common first NFA purchase. The rationale behind a 7.62N can is pretty simple: companies typically advertise them as being good for 5.56 and other smaller calibers as well.

But if you want to use your normal 7.62N silencer, undoubtedly not designed for AKs, there's some work you have to do to overcome the issues. Threads are often not concentric to the bore — blame Soviet precision. Rarely is an AK thread pattern readily available in your standard fare silencer, and there's no shoulder to speak of for a silencer to seat against. AKs are already so overgassed; adding backpressure from a silencer just makes it a heavily recoiling mess.

And yet, it can still be done. We'll show you a variety of ways to overcome these issues, all while using your normal 7.62N silencer. These methods can each be used piecemeal, or you can combine them. Before heading to your workbench, be forewarned that not all suppressor companies will honor their warranties if used on an AK, for all of the reasons we've already outlined.

ADAPTERS

First and foremost, we have to address the thread patterns on an AK. By far the most commonly seen thread pattern on a 7.62x39 AK is 14x1LH, though 24mm x 1.5mm takes second place, especially with 5.45 AKs. Other threading exists, like 22mm (for Romanian 5.45/5.56 guns) and 26mm specifically for some Yugoslavian AKs.

Regardless, your standard 7.62N silencer is unlikely to come in one of these thread patterns.

For best results, you should send your rifle in to be threaded properly in 5/8x34 — but not only can this be costly, it's also often impossible. So you're left with thread adapters.

But you can't just buy the first $15 thread adapter you find on Amazon. One major issue is that AKs don't have a solid shoulder to butt up against, instead utilizing a front sight block right at the end of the threads. Using adapters, in general and on AKs in particular, is a bad idea — you're stacking tolerances that you don't want to and exponentially increase your chance of a baffle strike.

If you have to use a thread adapter or a custom muzzle device for your patterned AK, use one that indexes on the muzzle itself and not on the front sight. We recommend the Griffin Armament thread adapter for general purposes, and not just some random one you found online cheap.

Take care to note that any specialized quick disconnect adapter threaded for an AK also should index on the muzzle, not the front sight.

If your thread adapter or muzzle device threads completely on, indexing on the front sight instead of the muzzle, it's a poor choice for use with a suppressor. If you just want a muzzle brake? Go with God. The main concern with suppressors is that any angular deviation from the bore is exponentially exacerbated the longer the device that you attach. A stubby 1-inch brake may give you no issue, but the same can't be said of a 6- or 10-inch silencer.

COST: $60 TO $70

The more parts you stack, the more you increase your odds of a baffle strike.

Always ensure your bore rod fits perfectly on the bolt face.

GAP (Good)

NO GAP (Bad)

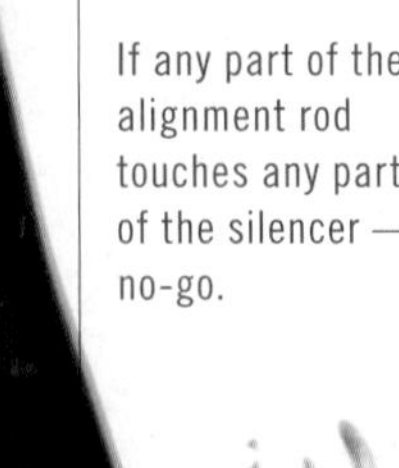

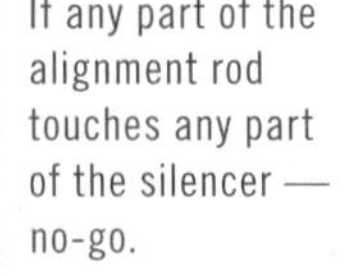

If any part of the alignment rod touches any part of the silencer — no-go.

RODS

Even if your chosen silencer has a thread pattern that works directly with your AK, you absolutely need to check alignment before firing to ensure you avoid a baffle strike and the shrapnel that can come along with it. With an AR-15 or similar, it's a fairly straightforward process to pull an upper and eyeball it, but the same can't be done with an AK. Even if you strip everything down, the rear trunnion will get in the way of a completely straight view down the muzzle.

So what can you do? Rod it.

Opinions vary on what alignment rod to use, with Geissele generally considered the best and CNC Warrior rods as a second. Some use precision drill rods from McMaster-Carr for the same purpose, though the tolerances aren't as precise as Geissele's offerings.

After mounting the silencer, insert the alignment rod into the end of the suppressor and allow it to fall to the bolt face of the rifle. If any point of the rod makes contact with the silencer — consider it a no-go and do not shoot the AK with the silencer attached.

If you're using either a 5.45 or 5.56 AK with a 7.62N silencer, you have a bit more wiggle room for alignment, but you still shouldn't pull the trigger on a rifle that fails an alignment test. AK-specific silencers, such as the Dead Air Wolverine PBS-1, use baffles with apertures that get continually larger further down the suppressor to help mitigate misalignment — but even still, don't shoot a gun that doesn't pass a rudimentary rod test.

COST: $19 TO $75

SPRINGING IT

As mentioned, AKs are gas hogs. Part of the mythical reliability of an AK-series rifle stems from the fact that they tap so much gas from the bore, allowing the bolt carrier group to fully cycle even when extremely fouled. Attaching a silencer to the end exacerbates this, with the backpressure the silencer creates pushing gas not only through the piston system, but also back through the barrel itself.

This leads to greater recoil and increased parts wear. No, regardless of what you've read on a gun forum someplace, the AK isn't infallible and can and will break.

Adding heavier springs doesn't solve the problem of excessive gas, but it does cut down on the symptoms. The addition of more robust springs will not only reduce recoil and muzzle rise, but will increase internal parts longevity. While there are several options, we've found the Snakehound Machine (SHM) spring kits to be fantastic for this purpose.

Not only do you receive a much stronger recoil spring in the pack, but also a hammer and extractor spring (the latter must be cut to fit). With most AKs, the SHM kits will work suppressed or unsuppressed, because the AK-series rifles have such a large operational window.

Swapping recoil springs on an AK may seem daunting at first glance, but it's actually a fairly easy process.

To replace your recoil spring:

1. Remove the recoil spring assembly.
2. Pull back on the spring and use a pair of vise grips to hold the recoil spring to the rear.
3. Spread the two spring bars to remove the end piece of the recoil spring assembly.
4. Remove the current recoil spring (Watch out! It can fly off).
5. Install the new SHM recoil spring and hold it to the rear with vise grips.

A pair of vise grips makes swapping in Snakehound springs a snap. Without them, you're liable to lose an eye.

6. Spread spring bars apart to reinstall end piece.
7. Remove vise grips and reinstall the parts.

Even if you choose another method for suppressing an AK, the SHM spring upgrade should be used in conjunction with it.

COST: $25

BUFFERS AND GAS BUSTERS

The increased gas from the addition of a silencer means your bolt speed will be higher than usual.

While a standard buffer setup will protect your rear trunnion from undue wear from bolt cycling, it also reduces some of the leaded gas rolling into your face. But you can take it a step further.

A common AR-15 mod is making a "DIY Gas Buster" charging handle utilizing RTV silicone, and the same can be done with an AK. For this mod, all you need is a cheap AK buffer, some cooking spray, and black RTV silicone (commonly available at any auto parts store).

While red RTV silicone is most commonly used on an AR, the black is more resistant to oil, so we utilize it for our purposes.

1. Field strip AK and degrease.
2. Install buffer on recoil spring assembly, install back into BCG.
3. Spray rear trunnion and inside of top cover with cooking spray to act as a release agent.
4. Gob the black RTV silicone onto the buffer and in/around rear trunnion.
5. Replace top cover.
6. Allow silicone to cure for 24 to 48 hours.
7. After curing, disassemble AK, remove excess cooking oil.

Ugly? Absolutely. Functional? You betcha. Gas to the face is significantly reduced.

COST: $18

GAS PISTONS

Every fix and solution up to this point only addresses the symptoms and not the root causes. But not this one. The KNS adjustable gas piston allows gas to vent from the piston itself, slowing the movement of the BCG. Less parts wear and reduced recoil — what's not to love? Well, a couple of things:

At the time of this writing, 19 different pistons are available covering the vast majority of AK/AKM variants. You'll need to choose the one that corresponds to your rifle.

In order to install the KNS gas piston, the original piston has to be removed. For many guns, this just involves a punch and a hand drill. But if you're unlucky enough to have a piston that's welded in place, like many CAI imported guns, installation is far more daunting. There's no guarantee that you won't destroy your bolt carrier in the process of removing a welded piston, and many don't consider it a worthwhile venture.

If you have a standard pinned carrier, removal of the OEM piston and installation of the KNS is fairly easy. All you need is a vise, hand drill, 1/8-inch punch, and 1/8-inch drill bit. OK, maybe a Dremel too.

1. Strip the BCG and put in a vise.
2. Locate the rivet head; sometimes you'll have to remove some finish to find it.

Black RTV Silicone being squished on as a seal against gas may look ugly — but it's effective. Not that the insides of AKs are pretty anyway ...

Hell, you probably have half of this already.

3. Partially drill the rivet out, and use the punch to completely remove.
4. Rotate out the OEM piston.
5. Install the KNS piston and secure with the included roll pin.

After installation, the piston has to be tuned. There are over 100 settings, so it may take you a bit of time to find your desired sweet spot. Also, bear in mind that the piston will have to be adjusted when taking the silencer on or off for optimal performance. Note that since the gas is vented internally, though carrier speeds are reduced, you'll still get the additional gas in your face.

COST: $149 TO $165

CUSTOM GAS BLOCKS

This option is for those more daring-minded, as it involves permanent modification to your rifle.

Unlike the KNS gas piston, this modification vents excess gas to the atmosphere, virtually eliminating any gas to the face. But in addition to permanent modification and tuning, you'll now have some small parts to keep track of.

1. Using a #21 drill bit (5/32 is the closest fractional size), drill a hole in the gas block forward of the end of the gas piston.
2. Thread the hole to 10/32, utilizing a quality tap and handle.
3. Obtain several vented cup-point cap screws — McMaster-Carr sells a 10-pack for $5.16 (PN 7573841421).

Now it's tuning time. Head to the range with the rifle and silencer of your choice, along with a set of numbered drill bits and a cordless drill. Since an AK is already so vastly overgassed, there's a decent chance it'll run unsuppressed with just the standard vented cap screw. If it doesn't, obtain an unvented cap screw from your local hardware store for unsuppressed use.

Install your suppressor and slowly open the hole in the vented cap screw until it reliably cycles at the speed you desire. While we'd normally advocate for a drill press, it's not practical for most to run to the range with them. A field expedient method is to use a pair of vise grips to secure the vented cap screw while drilling.

If you make the hole too large? Well, maybe that's why McMaster-Carr sells it in a 10-pack.

For our particular rifle/silencer combo we have one cap screw for suppressed, and another for unsuppressed. The extra screw, along with a hex key, is stored in a small baggy that conveniently fits into the AK cleaning kit.

COST: $10

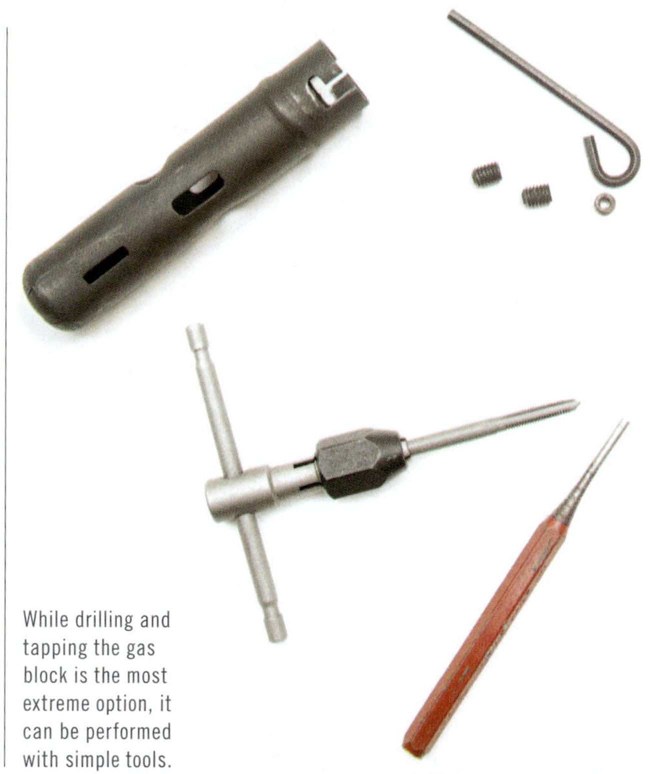

While drilling and tapping the gas block is the most extreme option, it can be performed with simple tools.

PILOTS, FROGMEN, AND SPIES'

The Knight's Armament XM-9, Snap-On Silencer

By Dave Merrill

Back when *The A-Team* and *Knight Rider* were still must-see TV, Knight's Armament was working on something a little different. The common name would be the Snap-On, but the official name when issued would be the XM-9. Ultimately, 3,800 KAC Snap-On cans were purchased by the U.S. Air Force for use for pilot escape and evasion kits in case they were downed in enemy territory, and also for Air Force Special Operations teams. An indeterminate number also went to the U.S. Army — and probably more than one clandestine service still has a handful in their inventory. While it's no longer the latest and greatest in silencer technology, it still holds a special place both in our hearts and in history.

DESIGN

The upgrade to the Frogman 1960s Hush Puppy, Snap-Ons were initially developed as a thread-on–type suppressor that we commonly see today. The second generation featured a locking gate-latch that was truly quick-attach when using a barrel machined specifically for this purpose. We're told this taper pin mechanism was the baby of then-KAC engineer John Anderson.

The design of the Snap-On is fairly rudimentary by modern standards, but remember, we currently stand on the shoulders of giants. The majority of the silencer itself is dedicated to a large gas expansion chamber, and then the projectile runs through a series of polyurethane wipes and spacers. As each bullet passes through the wipe, the wipe seals gases behind it — at least until they wear out. We've seen many configurations of Snap-On wipes, from simple small apertures to crosscuts, to straight-up make-your-own-hole-with-a-bullet. Quite honestly, the latter works very well.

The wipe and spacer assembly consists of seven (or eight!) PU wipes with two separate spacers. Wipes are certainly consumable items, and their longevity depends on the ammunition used and the rate of fire achieved. While one could just disassemble the Snap-On and repack with wipes after they wear out, the design of the Snap-On allows you to just swap out the wipe and spacer assembly right in the field if need be.

We're not entirely convinced it was a coincidence that the wipe size of the Gemtech Aurora, the successor to the Snap-On, shared the exact same dimensions. If you happen to own both, this is quite a boon. Let's just say we were more than happy to discover this.

SIGHTS

The dimensions of the Snap-On make it wide and squat, with a diameter bulging at 1.5 inches and sitting a hair below 5.5 inches in length. You're not going to be able to utilize modern suppressor-height sights very well, let alone those on a stock 92F with the Snap-On mounted. But KAC had a solution: Mount the sights directly to the silencer.

The top of the rear gate-latch mechanism serves as a rear sight, and a simple bead is threaded in place for use as a front sight.

SLIDE-LOCK

While this modified Beretta 92F wasn't anywhere close to being the only pistol at the time equipped with a slide-lock, it's probably among the last, if not *the* last. From our modern perspective, a slide-lock seems to only serve one purpose: to have a shot as quiet as possible — no action noise, no ejection port exhaust, no ejection port flame. However, they were initially developed for a different reason entirely.

Something that we take for granted with modern pistol silencers is the near universal use of a so-called Neilsen Device or muzzle booster. These recoil boosters were initially used on Vickers and Maxim machineguns in the early 20th century, but many years later found use in suppressed pistols. Since silencers add weight to the muzzle of a pistol, they can cause cycling malfunctions; a Neilsen Device temporarily uncouples the weight of the silencer when firing and allows for consistent, suppressed, semi-auto fire from a tilt-barrel action firearm. While the Beretta 92F doesn't utilize such a mechanism, many other pistols do. A slide-lock allowed for a quiet shot, as well as ensuring cycling malfunctions weren't present when using a silencer. Once Mickey Finn of Qual-A-Tec popularized the use of a Neilsen device with pistol suppressors, slide-locks went the way of the dinosaur.

In fact, in the early '90s, when the U.S. Government was developing the Mk23 Offensive Handgun Weapon System (OHWS), the first solicitation required a slide-lock to be incorporated. However, with silencer technology improving at a rapid pace and recoil boosters becoming more common, the updated government solicitation removed the slide-lock requirement entirely.

AT THE RANGE

Knight's Armament displayed their Snap-On with a box of Super Vel 147-grain subsonic 9mm, the ammunition

The majority of the Snap-On is dedicated to a gas expansion chamber.

Stacked sights definitely take some getting used to.

for which it was intended. Not having any of that on hand, we opted to use both Israeli 158-grain subsonic and 165-grain Freedom Munitions subsonic for our purposes.

The accuracy was, hmm, *rustic*. What you have to keep in mind is that each projectile is passing through more than three-quarters of an inch of polyurethane before it exits the silencer. Coupled with the fact that the rear sight literally moves every time the silencer is mounted, and you get some fairly inconsistent shifts. But let's be real here: This was never intended to be an Olympic target setup. Groups at 20 feet had significantly less shift than those at 30 but were still quite a bit to the right. Hey, this is one of the reasons wipe-only cans went out of style. That and the fact you can't use any sort of hollow point or expanding ammunition, unless for some reason you want a guaranteed endcap strike.

When utilizing supersonic ammunition, which absolutely degrades the wipes much faster, we hit consistently high.

Speaking of sights, we found the stacked sights to be rather optically confusing at first. When bringing the Beretta 92F equipped with a KAC Snap-On to bear, you're presented with two totally distinct sight pictures — the standard (blocked by the can), and then the ones you're supposed to use right on top.

What we were very pleasantly surprised with was the function of the slide-lock. As previously mentioned, the slide-lock allows for the quietest shot possible, because all action noise and ejection port exhaust is eliminated. After every shot with the slide-lock in place, it automatically dropped down from the slide and allowed for a very quick and efficient rack/eject/load. Some time was certainly spent with the design of this one.

LOOSE ROUNDS

In the year 2000, Advanced Armament Corp released their own version of the Snap-On. While it shared the same attachment mechanism, instead of a replaceable wipe pack, they rocked standard baffles. It was also considerably

With supersonics, we were consistently high at 20 feet, and consistently right with subs. Neither grouped very well.

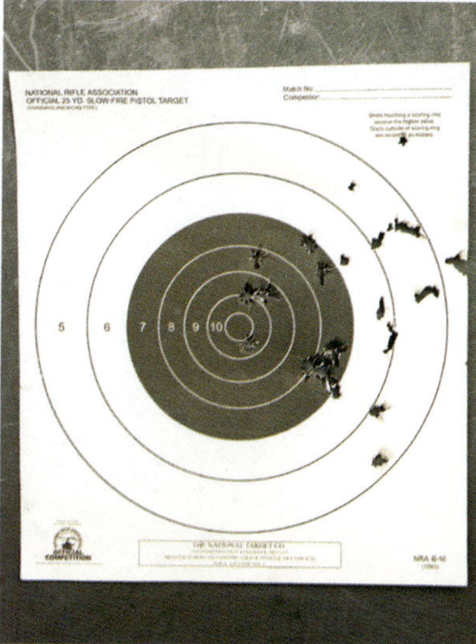

The ambidextrous slide-lock is very easy to lock and unlock, even with gloved hands.

The AAC version of the Snap-On was skinnier, but considerably longer than the original.

longer than the Knight's version. Trade-offs with everything.

While the Knight's Armament Snap-On is far from a modern-day pistol suppressor, it holds an important place in the history of silencer development. Looking back, you can't help but feel like you're looking at photos of yourself back in middle school; You can recognize yourself, but you dressed really awkwardly.

Knight's broke out their old tooling (or maybe checked a dusty warehouse) and offered up 188 silencer/barrel combos for $1,999.95 each, and a mere 19 full kits as you see here for $8,684.12 a pop. Despite the Snap-On and Beretta slide-lock being dated, there are very few that wouldn't be more than proud to have one in their collection. R

Below you see the contents of the entire kit, ammo and all.

QUIET PRECISION

What You Think Matters May Not Be True

by Dave Merrill

Sure, one could just put whatever 30-caliber silencer on the end of a long range rig, but that's not what this article is about.. At the base level, it seems that the main use of a suppressor is to well, suppress sound. However, when it comes to long range specifically, you may find sound suppression to be further down on the list than you'd expect.

Ryan Hey, professional Precision Rifle Series (PRS) shooter and MagnetoSpeed rep broke it down for us like this:

"The first thing I care about is accuracy. If a silencer decreases accuracy, it's gotta go. Next up is repeatability. Can I take it off, then put it back on without any major degradation? Reduction in recoil is next on the list; I have to be able to spot my own hits at longer distances. Last up is sound. If I don't need to wear earpro, all the better."

Yes, sound is the very last thing Hey cares about. He still cares about it—it just isn't the first priority. Some of your vanilla 30-caliber cans will do some of those things, but a quality long-range silencer will do *all* of those things. Remember that these are generalities, though we make some recommendations at the end of the article.

RECOIL AND SIGHT TRACKING

One reason we have seen some PRS and long range shooters switch back to muzzle brakes from silencers over the past couple of years is due to the more effective recoil control some brakes offer. The ability to see and spot your own impact is vitally important to long-range success.

Not every silencer will reduce your recoil, but we found Leviathan Suppressors particularly excelled in this role. Using both 6 Creedmoor and 260 Remington we were able to spot our own hits from 400 to 900 yards with no issue. Let's just say that we didn't want to take the Leviathan off our rifle at the end of the day.

POINT OF IMPACT SHIFT

We've come across many claims that XYZ brand silencer has, "zero point of impact

The hybrid design of the Crux brake offers excellent repeatability.

shift.". If there were actual truth in advertising the claim would be more along the lines of, 'POI shift is small enough you probably won't notice. Probably. Maybe.'

Many factors come into play when we're talking about POI shift with a silencer on vs. silencer off. The weight of the silencer, the length and profile of the barrel, the change in harmonics, and the shift in velocity allow add up to at least some amount of shift.

There's going to be some shift—what we really care about is if the shift is repeatable. In an ideal world, you'd re-zero every time you took a silencer on or off, but we don't live in a world of strawberry rainbows and marshmallow clouds. Our best advice is to never take a silencer off, but if you think you'll be hot-swapping, take note of any shift.

TO QD OR NOT? THAT IS THE QUESTION

The argument between direct-thread (DT) and quick-disconnect (QD) for silencers in general has long been the subject of debate; and it's no different regarding long range shooting.

While a QD allows a user to rapidly remove a silencer (usually for transportation), not all QD systems are created equally. For example, the bi-lock mount used by Gemtech, GSL, CMMG, and NEMO reciprocates on the mount when fired—perfectly fine for most AR-15 shooters, but it doesn't aid those shooting long range. Currently the most repeatable locking QD mounts are those that only mount one way and don't rely on ratcheting teeth or similar mechanisms on the muzzle device itself to ensure they lock in place. Standouts here include SureFire and Dead Air.

Direct-Thread mounting should be considered a semi-permanent affair; while you can take it off, a direct-threaded can is really meant to live on the muzzle most of the time sans some cleaning or maintenance. As such, this option is more popular for those with folding stocks and chassis systems, because, generally, they can be transported with a suppressor mounted. DT also usually comes with a lower price tag, and more internal volume because there's no muzzle device taking up space in the blast baffle.

But, there is a third option. We call it a hybrid, though some others refer to it as a non-locking QD. With a hybrid system, the silencer is still threaded on, but on a specialized muzzle device rather than the barrel itself. The best ones have a taper mount, which actually align a suppressor better than their DT brethren. The advantage of these systems is that you're able to remove a muzzle device (albeit with more time than a QD) while retaining decent repeatability.

BLAST MITIGATION

If you shoot a brake very close to the ground, you're likely to get dust, rocks, and dirt thrown everywhere. The same can't be said of a rifle equipped with a silencer. Since most of the blast is contained inside the suppressor itself, there isn't much energy left over to kick up debris around the muzzle. This makes a shooter's position much harder to track from an observer's or target's position.

This is particularly useful for military and police snipers, but really no one wants all that sh*t thrown in their face regardless of their role.

SOUND

No matter how you cut it, if you're shooting a precision rifle at any sort of range it's gonna have a supersonic crack. This isn't to say that the effective decibels do not matter, just that they generally fall below everything else on this list of priorities for some long-range shooters. R

Leviathan makes silencers in several sizes.

AREA 419 MUZZLE DEVICES

Something we hate, but silencer manufacturers absolutely love, are proprietary muzzle devices — and it's not hard to see why. If you buy a silencer from Company X, chances are you'll buy multiple muzzle devices also from Company X (most at more than $100 each), so you can use it on multiple guns. When it comes time to purchasing silencer #2, just who do you think you'll more likely be buying a silencer from?

It's like voluntary DRM for your guns.

A company looking to help people break out of this is Ohio-based Area 419. While they're not well known outside of the precision crowd, they should be.

Their patent-pending Hellfire modular system doesn't have the same limits as other proprietary mounts. This is one of the so-called "hybrid" systems we talked about.

First and foremost, you can use the easy-to-align Hellfire brake. But it only gets better from there.

All with the same Hellfire muzzle device, you can get adapters for a number of precision rifle silencers.

One muzzle attachment gives you the ability to directly attach a Hellfire brake, and, through the use of Area 419's suppressor mounts you can attach TBAC, Crux, SAS Tomb, SilencerCo Omega (and therefore Dead Air Nomad), SilencerCo Harvester, Saker, and universal 5/8-24 direct threaded suppressors — with more coming.

Area 419 offers the same functionality for calibers up to .375 with its Sidewinder muzzle attachment system.

MAKE:
Area 419

MODEL:
Hellfire Universal Adapter

MSRP:
$40 plus $55-$165 for silencer adapters or Hellfire brake

URL:
area419.com

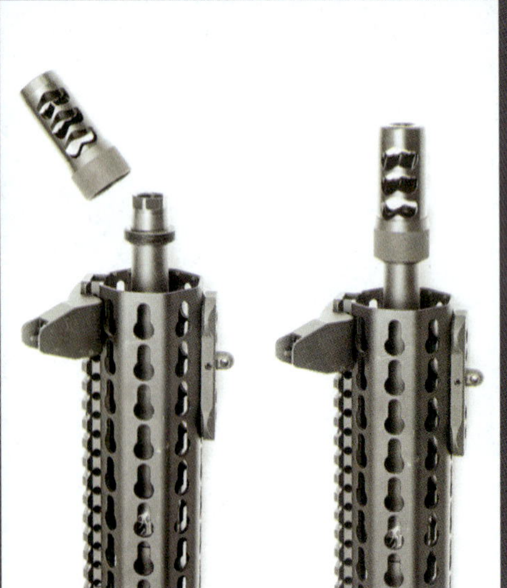

SUGGESTED SILENCERS

LEVIATHAN
30 CAL REAPER

MSRP
$1,195

URL
LeviathanSuppressors.com

NOTES
Of all the long-range silencers we used, the Leviathan cans had the best recoil mitigation. There's no free lunch, however. Using the Leviathan on a gas gun, which they specifically don't recommend, had the opposite effect due to increased back pressure. Keep this one on a bolty and watch how little your reticle moves.

CRUX
ARK NEO

MSRP
$1,103

URL
CruxSuppressors.com

NOTES
The Crux (pronounced "crow") Ark Neo has proven to be a versatile titanium little beast. Working well on both gas and bolt guns, what impressed us the most was the repeatability using both the factory Precision Mounting System (PMS) and the aftermarket Area 419 adapter.

SUREFIRE
SOCOM260-TI

MSRP
$1,349

URL
SureFire.com

NOTES
We're pretty sure someone had to hold SureFire's Barry Dueck at gunpoint to get him to design a silencer that won't live on a belt-fed, but he did it. The SOCOM260-Ti is the first silencer from SureFire entirely designed with accuracy in mind.

Trijicon
RM01 2PE1:19
MADE IN USA
154124
GREY
XVL2
SUREFIRE
LASER RADIATION: AVOID DIRECT BEAM EXPOSURE
VISIBLE AND INVISIBLE RADIATION IS EMITTED FROM
GREY GHOST PRECISION
MADE IN USA LAKEWOOD, WA
GGP

EVOLUTION, NOT REVOLUTION

SureFire's Latest Silencer Update: The Ryder 9-Ti2

By Dave Merrill

If you like your SureFire, you can keep your SureFire — at least that's what the employees of the Fountain Valley, California, company tell us. Fresh in the booth at SHOT Show 2019 will be a brand-new 9mm silencer from SureFire. The aptly named Ryder 9-Ti2 is the next step for this popular pistol can.

While it seems the pistol suppressor world is currently circling their wagons around modular 45ACP models, SureFire isn't going down that road. It's not hard to see why the industry has been hot to trot for 45ACP silencers; when they're designed well they can be used on nearly every lower caliber. Gone are the days where a 40-caliber silencer is on many company SKUs. With that said, there are benefits to a suppressor specifically made for a given caliber, which is also why 5.56 and similar dedicated caliber cans continue to exist to this day. The same goes for the Ryder 9-Ti2.

No, instead of coming to a whole new design, they've decided to spend time giving their current line further refinement.

We'll give you the bottom line up front: If you currently own a Ryder 9-Ti, you don't need to run out and upgrade right now. But if you don't own one and are

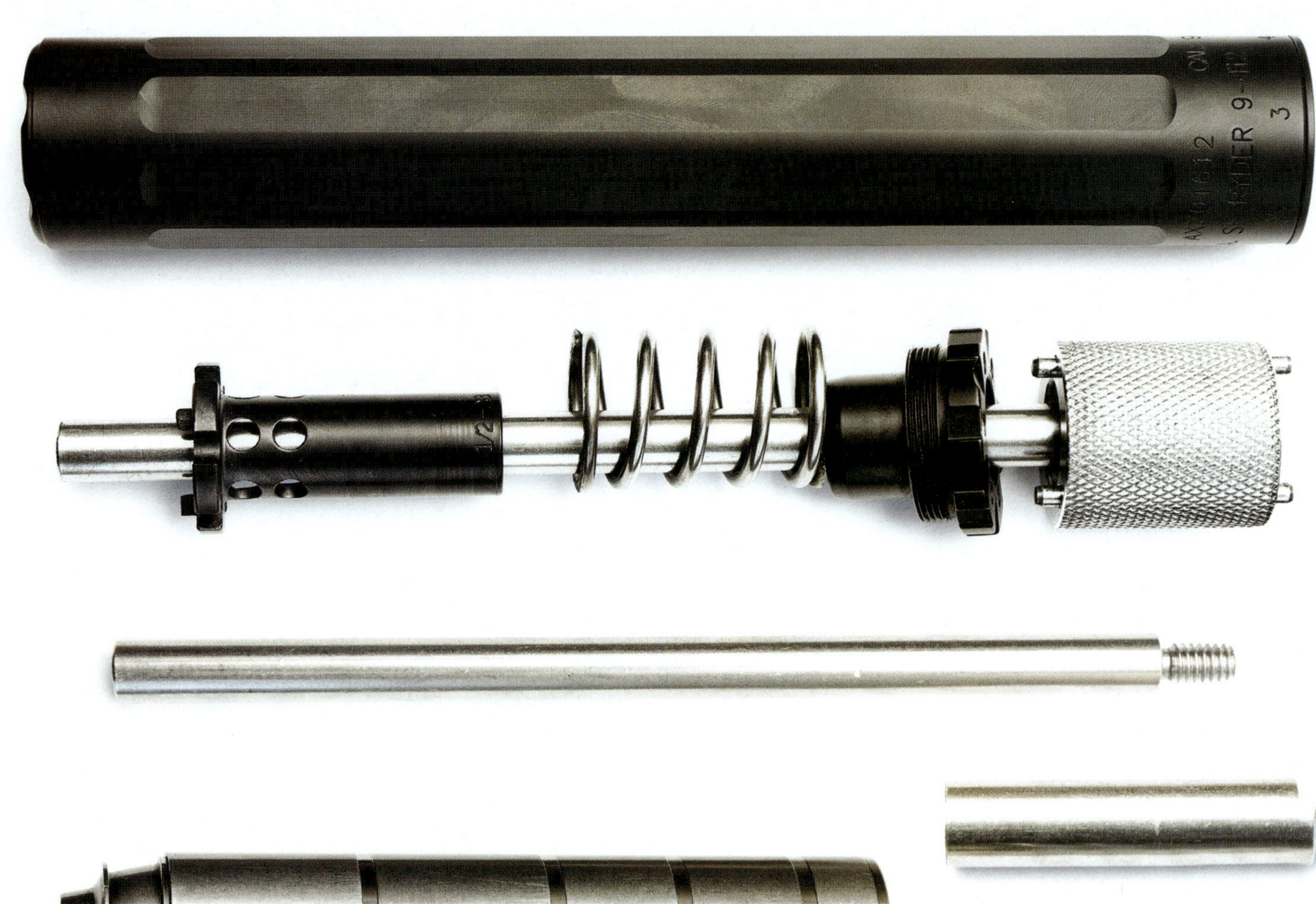

considering a 9mm silencer, give this one a hard look.

The Ryder 9-Ti2 comes with all the tools you need and then some.

SO WHAT'S THE DIFFERENCE?

There are far more similarities than differences between these two generations, hence why this is a sequel and not an entirely new franchise. There's the same performance, slim 1.25-inch diameter, and curves. To fully answer the question of revisions we'll give you two words: strength and longevity. You won't spot most of these differences yourself without some side-by-sides and some guidance.

The Series 2 model has an improved interface with the pistol at the rear. This upgraded mechanism has longer mating and bearing surfaces for greater strength and more consistent alignment. Secondly, the booster assembly has been beefed up and a stronger spring is used.

This comes at a price not of more dollars but of additional weight. The Ryder 9-Ti2 is two ounces heavier, at 11.5 ounces, than its predecessor.

These changes weren't really made with Americans in mind, but instead to ensure foreign entities issued the Ryder 9 had more time between maintenance intervals. If a silencer goes down for an American consumer, SureFire is just a quick post office trip away. If a silencer goes down for a mil/LEO unit, they can dig into their spare parts and pieces knowing they can get catalog parts replacement in mere days. But a military contract in Myanmar? (Note: We have no idea if SureFire is used in Myanmar, but we liked the way it sounded.) Even airmail takes some time to cross the Pacific.

But just because these upgrades weren't made for us, we still benefit from them. At the time of this writing, we're told SureFire will only be producing the Ryder 9-Ti2 and letting the old ones go. Just think of it like gently shushing the person you're holding the pillow over.

BREAKING IT DOWN

If you're not familiar with the Ryder series or pistol-caliber SureFire cans, this next section is for you. We'll break down the design and point out features that we found very thoughtful on the part of the engineers.

Each Ryder 9-Ti2 ships with tools. This shouldn't be a huge deal, except for the fact that more than once we've been forced to buy a special $50 tool simply to remove a suppressor endcap. The three-piece set consists of a threaded rod, a special adapter to push out baffles, and a knurled cylinder wrench with stems on either side for end cap/front cap removal. The reason for stems on both sides? So if one breaks, you can simply flip it over. It's the little things.

To disassemble the Ryder 9-Ti2, thread the rod into the knurled cylinder, insert the rod into the bore, and turn counterclockwise; this works for either end of the silencer. When the endcap is removed, slide the baffle removal tool over the tip of the rod, insert, and push. These silencers are made with very tight tolerances, so once you get a prudent amount of projectiles down the pipe they can be hard to take apart. We note that SureFire recommends giving the booster a scrub every 300 rounds and the baffles every 1,000, though we regularly exceed these figures purely out of neglect and laziness.

There are a total of six baffles inside. Each baffle has a slightly different length, so there are no spacers sans the one immediately in front of the booster assembly. They're (thankfully) clearly numbered in the order of assembly.

After you've done your share of cleaning and scraping, assembly is up next. However, don't simply click the baffles in place and let it ride, lest something get misaligned. Slide your baffle stack and end plate directly onto the tool, and then use the entire apparatus to lock it down all at once. There's a pleasant ratcheting when you get toward the end.

SUREFIRE
RYDER 9-TI2

CALIBER
9mm

WEIGHT
11.5 ounces

LENGTH
7.8 inches

DIAMETER
1.25 inches

MSRP
$849

URL
SureFire.com

AT THE RANGE

Regarding sound, we would absolutely consider the Ryder 9-Ti2 "hearing safe" indoors with subsonic ammunition and outdoors with supers — though OSHA would certainly disagree with us. If you've heard a Ryder 9 already, this one will be the same. That is to say: excellent performance with great tone.

LOOSE ROUNDS

The SureFire Ryder 9-Ti2 isn't revolution; it's evolution. In a world of modular 45ACP silencers, the Ryder still holds its own. Besides, it's not like you're still shooting 40-cal anyway. R

Let's face it: Integral silencers are cool as hell. Who doesn't want an OSS HD-Military, the (mostly) silent pistol used to dispatch Nazi officers? No real Americans, that's for sure. But are they worth it? Many different integrals have graced the pages of RECOIL in the past, but this piece isn't about a particular integral gun; we're going to talk integrals in general.

THE MANUFACTURER'S VIEW

As far as integrals, experts in the silencer community are somewhat split. Evan Green of Griffin Armament told us:

"Integrals have long caught the attention of boys young and old alike. The large, long, suppressors purposely built into the host firearm, beckon images of the French resistance in World War II. Appearance is 99 percent of reality, right? Wrong. Bigger isn't always better. The problem? Most integrals have never been researched and designed with a large enough monetary investment to yield high performance. This is because most modern consumers want versatile suppressors they can use on a collection of firearms as opposed to a single one."

FULLY INTEGRATED

We Let You Know When an Integral Silencer is Worth It

By Dave Merrill

Todd McGee of Dead Air has a slightly different view:

"A big positive is that it allows a weapon system to be optimized for a given use. Many people don't stop and think of the host weapon, ammo, and suppressor as system. Any one of those factors has a huge impact on the overall performance, not just in terms of sound at the muzzle, but also in reliability, precision and accuracy, and sound at the shooter's ear.

"Obtaining a single-minded focus on something always comes at a cost. Just like an athlete who specializes in one area (say, body building) they're giving up on something else (say, fitting into skinny jeans)."

AMERICAN EXCEPTIONS

NFA laws themselves are a large part of the reason why integrals aren't more popular. But this is also the reason why hoodrat integrals exist as well. Simply put, a hoodrat integral (lovingly called, "fake-ass integrals" by some of the staff) is a normal silencer permanently attached to a rifle in order to bring the barrel

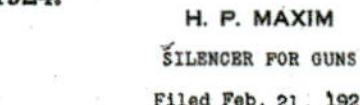

U.S. Patent Jul. 24, 2018 Sheet 2 of 6 US 10,030,929 B1

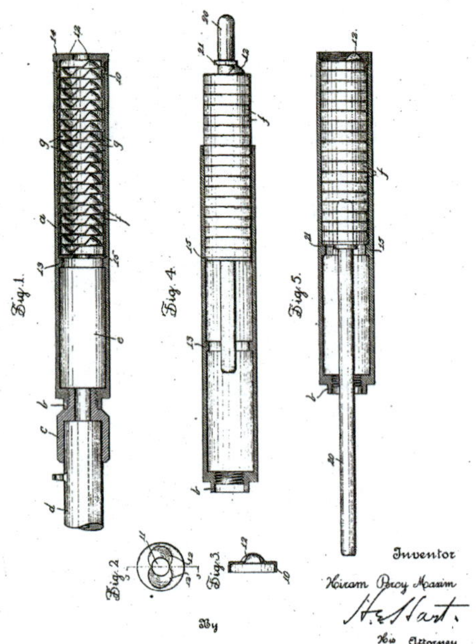

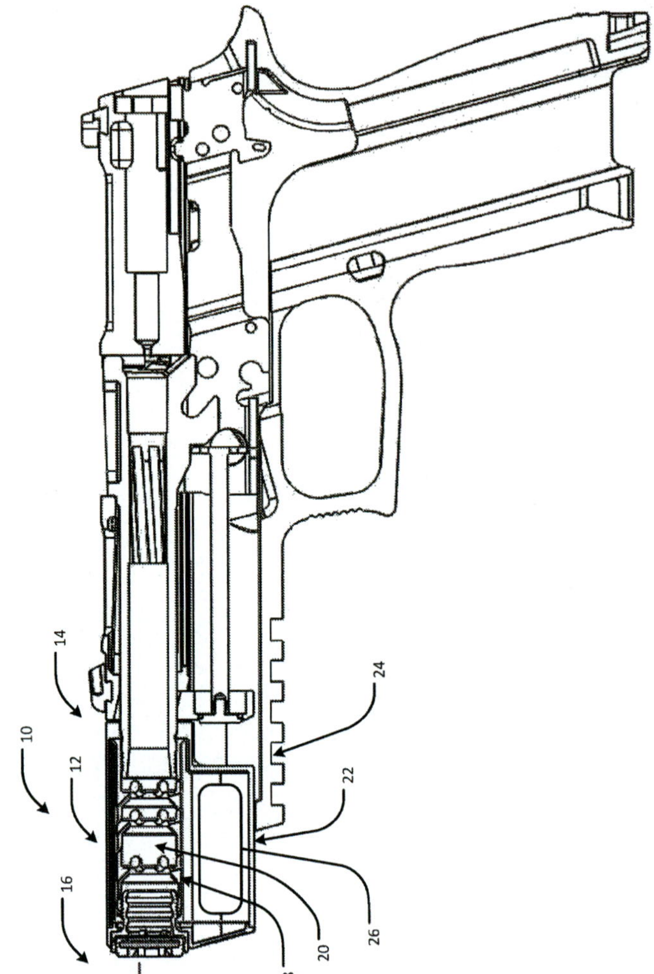

length to the legal minimum of 16 inches. Not only does this make a rifle a "single stamper" because only the silencer tax stamp has to be paid, it also allows interstate travel without pre-approval from the BATFE.

For some reason, likely to do with politicians not understanding the laws they're passing in the first place, or heavy drinking, a Form 5320.20 has to be filled out, submitted, and accepted before you cross state lines with a machine gun, short-barreled rifle, or short-barreled shotgun — but not with a silencer or an AOW; those you can just roll with, provided they're legal where you're going.

With a normal suppressed SBR, you'd have to fill out some forms well in advance of leaving the state, but it's not the case with a hoodrat integral. That's virtually the only upside, because while the silencer is permanently attached, you're not getting some of the benefits an actual integral can provide.

The hoodrat integral is potentially negated with the recent uptick in totally functional and usable pistol "braces." While the legality of shouldering them can change on the mood of someone in the BATFE Examination Branch, currently it's A-OK.

If you still want to pursue a hoodrat integral, bear in mind, oftentimes you're going to be stuck with whatever gas block and handguard you currently have once it goes on — at least if you want to do it yourself. Any decent SOT will be able to remove it and replace it, but that's a lot of cost to add for what would otherwise be a simple handguard swap.

Many silencers, such as this one from Odin Works, come with a hole all setup for blind pinning.

Some of the guts from actual integrals.

THE CASE FOR A MODERN INTEGRAL

Legal reasons aside, what McGee pointed out about a truly integrated system still rings true. Just like how we all benefited from government requests for a new handgun, Uncle Sam looking for a Suppressed Upper Receiver Group (SURG) gifted the public with a bevy of options. The Gemtech Integra, as seen in RECOIL #35, is the first one that comes to mind. With their patented bore evacuator, they took a high-pressure 5.56mm round and made it shoot soft and quietly. Along similar lines, the Liberty Suppressors integral 300 BLK, the Leonidas, remains a quiet option.

The added volume for gas expansion an integral provides can also be beneficial, just not as much as you imagine. The explosives gases from a fired shot absolutely will travel backward (evidenced by that gas-in-your-face with many systems), but it takes more effort than one would imagine. That extra volume for gas expansion can also result in a louder First Round Pop (FRP), which is one of the reasons monocore designs are generally louder than their traditionally baffled brethren.

LOOSE ROUNDS

We've often said integrally silenced weapons are absolutely not for first-time NFA tax stamp owners, and we continue that same advice to this day. One thing remains for certain: An integral silencer will absolutely always be cooler than one that attaches, but the juice may not be worth the squeeze when you consider the American tax stamp system. If silencers are somehow magically removed from NFA — fat chance with President Cheeto exclaiming, "I don't like them at all" — there would be about zero reason for everything to not be an integral.

With our current legal system, the integral remains a super-cool, but superfluous purchase for many. R

INTEGRAL RPR

Witt Machine

By Dave Merrill

No, you haven't traveled through a time machine where John Arthur Ceiner cans are actually decent. Thankfully, there's been no more of that in recent years. Nor have you come across some sort of April Fools' joke, where we are reviewing and advocating something from the same realm. Mitch WerBell? Oh hell no, our major throwback issue was the last one, not this one.

The integral RPR from Witt Machine only looks goofy if you don't have any reference point. Namely, it only looks a bit dickerty, because it's sleeved from the muzzle all the way down to just forward of the chamber.

Inside the guts, there's a different story to be told. While usually we'd want an integral gun to be shorter than a barrel/silencer combination, since the RPR is built for long range, shortening the barrel would effectively shorten capabilities. And since this one happens to be in 308 Winchester, we want all the capability we can get.

DESIGN

Witt Machine starts with a basic 20-inch RPR barrel, then a 10-inch two-piece monocore is threaded on (the first section being a brake-style and the second being more of a traditional monocore).

The exterior of the barrel is threaded near the base to accept the rear sleeve retainer. Then just below the 1.75-inch diameter sleeve rotates onto the whole shebang. Six hex screws are used to secure the end of the sleeve to the monocore, also giving the added benefit of adding some rigidity to the barrel itself. The Master of Arms Enyo, featured in Issue 33 used a similar method to increase rigidity.

Still, at just over 48 inches long with the stock unfolded we aren't talking about a tiny titan.

... WELL WHY BOTHER?

The Witt Machine Integral RPR isn't smaller than a regular RPR with a silencer, but with the folding stock it's even smaller than our 16-inch barreled Grey Ghost Precision rifle and Crux ARK NEO silencer — and that's a rather compact getup.

And there's one huge benefit to that John Holmes sleeve: an absolute ton of gas expansion can take place aiding in how quiet this bolt gun really is.

The sleeve allows for a helluva lot of gas expansion.

SOUND

Unless you're shooting subsonic 308 Win (if you want that — buy 300 BLK) there's always going to be the crack of the projectile as it breaks the sound barrier. So far we've heard dubious reports of decibel ratings, and lacking the Brüel & Kjær pulse system and microphone which costs roughly $35,000 we just have to go with our qualitative perceptions. No, your Radio Shack or Amazon decibel meter isn't suitable for the task.

Since the RPR is a bolt action, there's absolutely no ejection port noise, and since the end of the barrel is roughly 40 inches from your ears? Yes. Yes, it's very quiet. We'll take a stab in the dark and say somewhere in the 130s — either way, very comfortable to shoot outdoors without ear pro.

The monocore starts with a brake pattern before decisively moving to a more traditional design.

With its folding stock, the Witt Integral RPR is no longer than a 16-inch rifle with a can and collapsed stock.

KITTING OUT

We equipped the Witt Machine RPR with an Atlas BT46-LW17 PSR 6-9-inch bipod and topped it off with the exceptional Bushnell XRSII 4.5-30 scope in a ZRODelta DLOC-M4 34mm mount. While a 30-power might be more magnification than a 308 Win requires, we'd much rather have too much magnification on the high end rather than the reverse. The addition of a Lead Faucet Tactical Brokos brace made it much easier to toss around on both the range and in the bush.

For ammunition we decided on Hornady's 308 Win 168 grain ELD Match — a known performer.

AT THE RANGE

Transporting the Witt Machine RPR wasn't too much of a burden thanks to the folding stock. It easily fit into a Pelican iM3200 Storm case with some room to spare, and we even managed to squeeze one into the much smaller Grey Ghost Gear Rifle Case — though it was akin to stuffing Mama June into size small yoga pants. Accuracy was directly on par with what you'd expect from an RPR; we managed to squeeze a 0.8-inch 10-shot group out of it at 100 yards, but we know that this rifle is more capable than we are.

You have to use a handguard with a large enough inner diameter to accommodate the suppressor.

Targets out to 900 yards were easy to smack, and our main limitation was the caliber itself rather than the platform or silencer. We had little issue with spotting our own shots like shooters have to in PRS and other long-range matches.

BITCHES & QUIRKS

Though the RPR can accept any standard AR-15 handguard, there are some limitations when it comes to this particular rifle. First and foremost, not every handguard has a sufficient inner diameter to accommodate the Witt Machine integral barrel. Additionally, many AR handguards have a goofy gap when mounted on the RPR just by the nature of how it attaches.

Our model features a Seekins Precision MCSR V2 Rail System equipped with M-LOK attachments. We've also seen Witt Machine integrals sporting PHNX HexGuard handguards from VDC Armory. This isn't the first time we've seen this particular Seekins handguard used with an integral; you may recall the Gemtech Integra featured in Issue 35 used the very same rail, and likely for the same reasons.

But don't just think you can simply attach any M-LOK accessory willy-

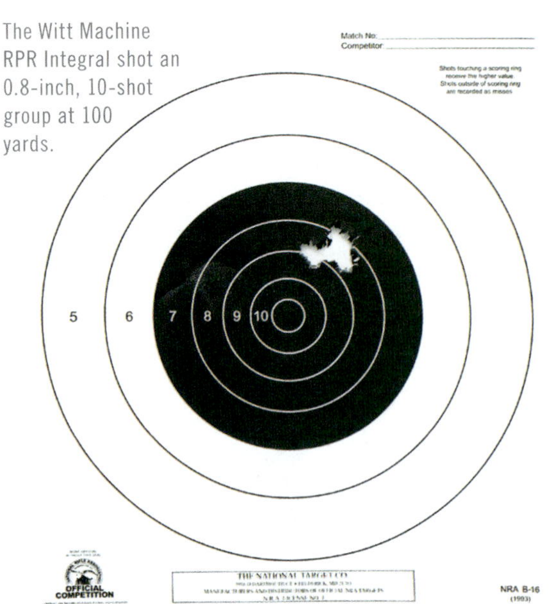

The Witt Machine RPR Integral shot an 0.8-inch, 10-shot group at 100 yards.

nilly. Just like the aforementioned Gemtech Integra, the screws have to be shortened to prevent damaging the outer sleeve of the silencer. A couple minutes with a hacksaw and file and we were in business, but it's definitely something you have to keep in mind. Regardless, you're absolutely going to want to plan your accessory positioning beforehand. We found it far easier to just completely remove the handguard to install parts à la 2008 with a Troy rail.

Our main complaint wasn't really the noise, accuracy, or length (once we got some perspective, at least), but about the caliber itself. However, fear not, Witt Machine produces these bad bears in many calibers — we just lost at the roulette wheel and ended up with a 308 Win.

LOOSE ROUNDS

While the Witt Machine Integral RPR isn't much smaller than a standard rifle and suppressor, it certainly is much quieter than your average fare. This same performance with a 2006 M4-2000 on a rifle you will not get. While we can't comment on free-bore boost, it certainly doesn't apply with a silencer that's totally integral. As with every other integral, this isn't a first NFA purchase for basically everyone, but if you're a dedicated long-range shooter, this one has a place.

There are some quirks, such as having to slightly reduce the length of the M-LOK screws lest you molest the outer tube, but these are easily overcome with a little bit of common sense.

As stated before, if the RPR didn't have a folding stock, this one would be a no-go. But as it stands? Buy with confidence. You're welcome. R

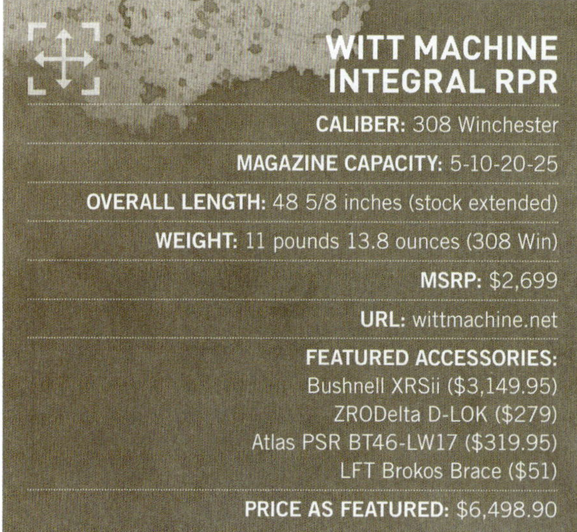
WITT MACHINE INTEGRAL RPR

CALIBER: 308 Winchester

MAGAZINE CAPACITY: 5-10-20-25

OVERALL LENGTH: 48 5/8 inches (stock extended)

WEIGHT: 11 pounds 13.8 ounces (308 Win)

MSRP: $2,699

URL: wittmachine.net

FEATURED ACCESSORIES:
Bushnell XRSii ($3,149.95)
ZRODelta D-LOK ($279)
Atlas PSR BT46-LW17 ($319.95)
LFT Brokos Brace ($51)

PRICE AS FEATURED: $6,498.90

TI FIGHTER

GSL Technology SWAT-5 Silencer

By Alexander Crown

By our count, there's somewhere around 40 individual silencer companies in the United States. That makes for an incredible amount of options for consumers when choosing a silencer. So, what factors should go into that decision? Price, length, weight, sound reduction, features, accessories, warranty? Absolutely, every single one of those should play a part in what can you buy. Another factor could be lineage. Silencers have been around over 100 years now and many notable inventors and companies have come and gone. One company that has hung around for over 30 years is GSL Technologies Inc.

The name may sound familiar as they were universally known for being associated with Gemtech up until a few years ago. Now they're on their own, and the wizard behind the curtain is Greg Latka. Latka has a history of manufacturing, particularly silencers, and holds numerous patents for several designs. His old, but at the same time new, company is cranking out cans for all consumer interests from .458 SOCOM down to rimfire. We were sent a 5.56mm specific, direct thread model, adequately named the SWAT-5.

THE CAN

The SWAT-5 is a direct thread silencer, with the industry standard ½x28mm TPI pattern. The outside diameter is 1.5 inches, typical of most modern rifle silencers, and the length is 6.3 inches. Again, similar to most modern rifle silencers. The materials for the SWAT-5 is 6AL4V Titanium, for the tube and we're guessing (based from intimate product knowledge of similar silencers) that all baffles with the exception of the blast baffle are also titanium. GSL lists the other material as Inconel, which is a super alloy that holds up well to erosion

The SWAT-5 looks right at home on a short-barreled rifle.

making it an excellent choice for a blast baffle for 5.56mm and short barrels. Our educated guess was that the SWAT-5 is a four M-Baffle design, with the entrance chamber making up about half of the internal volume. X-rays proved us right.

These two materials are found in numerous other silencer designs and help with keeping overall weight down while increasing total durability. The listed weight for the SWAT-5 is 13 ounces. Being a sealed silencer means màintenance comes at a minimum, however GSL does advise to keep your retention device components clean.

Direct thread silencer designs are as simple as it gets. Screw them onto the end of your barrel and shoot. The main benefit is there's no need to worry about extra mounts, muzzle devices, or washers and shims. Generally speaking, when you screw a silencer on a barrel you should check alignment, and when possible, torque it in place, possibly even apply a thread locker to the barrel threads. Many designs have wrench flats or other external support to aid in the installation but the SWAT-5 doesn't. GSL engineered a built-in silencer retention system that uses a washer and a locking collar — the washer is concave and labeled to prevent screwing up the installation. While a properly mounted and torqued, direct thread silencer won't come loose, the GSL device makes this attachment method idiot proof with a simple, patented tension device.

The manual (which we did read) states that the washer is designed to free float and movement isn't a concern. When asked about the retention device, here's what Latka had to say:

"I developed that system sometime in 2012 and filed a patent. It was in response to a solicitation that we lost because the suppressors would become

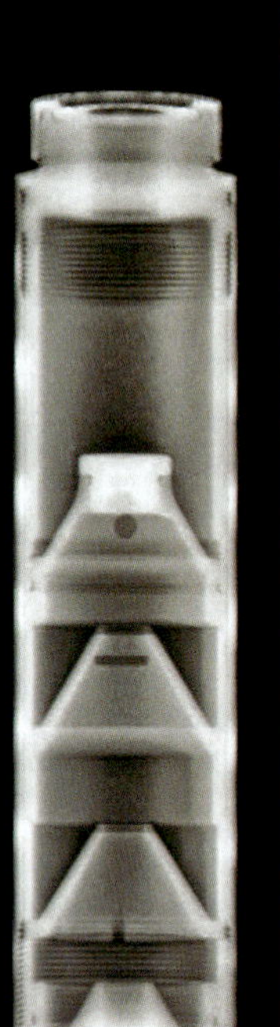

unthreaded during long periods of sustained fire. That may have been from the testing folks not knowing what they were doing, but either way, I used this simple system to ensure that cans wouldn't come loose when subjected to that type of shooting. As the name suggests, the SWAT-5 is meant for entry teams or other officers that may have a need for high round counts and do not want to worry about their can coming loose."

AT THE RANGE

The SWAT-5 has a minimum barrel length of 10.25 inches, so naturally we put it on a 10.5-inch machine gun. Accounting still hasn't approved our $30K request for the latest sound testing equipment and OSHA won't return our calls, so we rely on our silencer experienced ears to determine sound reduction. Suppressing the blast on a 10.5 inches may seem like an easy feat at this stage in silencer technology, but not everyone can do it well. GSL tamed the blast with ease and sound levels were comfortable. Advertised sound reduction is 30 dB and doesn't state what barrel length that's measured on or where the sound meter is located. We think 30dB is a stretch for a 10.5 inches, but completely obtainable on a 16-inch barrel.

Another subjective measurement with silencers is the amount of gas blowback experienced. The machine gun ran flawlessly with American Eagle .223 ammunition and gas was blindingly uncomfortable. This may be attributed to the larger entrance chamber on the SWAT-5 because we were not using any adjustable bolt carriers or gas blocks.

During twilight hours and into the darkness the SWAT-5 did a decent job at signature reduction. Flash from a 10.5-inch barrel is excessive at night even with a flash hider and this silencer took it down to very little. This helped keep natural night vision from getting wrecked every time we sent lead down range.

GSL TECHNOLOGY INC. SWAT-5

CALIBER:
5.56x45mm

CONSTRUCTION:
Welded Titanium and Inconel

LENGTH:
6.3 inches

DIAMETER:
1.5 inches

WEIGHT:
13 ounces

DECIBEL REDUCTIONS:
30 dB

MSRP:
$850

URL:
gsltechnology.com

CONCLUSION

The SWAT-5 from GSL Technology is a robust direct thread silencer. The company's reputation with manufacturing silencers translates to a quality product that'll be at home for Law Enforcement or the average gun enthusiast. The simple attachment method and the patented silencer retention device mean this silencer is probably best served on a dedicated gun like a duty gun. R

DIY SUPPRESSOR

Here's How to Legally Make Your Own Can, Save Yourself a Year of Wait Time, and a Few Hundred Bucks

By Iain Harrison
Photos by Kenda Lenseigne

Making your own firearm from a partially completed receiver is something we've covered extensively in the past, but there's a perfectly legal route to achieving the same ends with a suppressor, which in many ways is more effective and attractive for the average law-abiding citizen. In order to transfer a factory-made can from your local dealer, you'll have to fill out an ATF Form 4, pay a $200 tax, submit fingerprints and a passport photo, and then wait. And wait.

After around 11 months, you might get the chance to go pick up the property you paid for so long ago, or you might just have to wait a bit longer. If this doesn't sound quite so appealing, then you could always go the DIY route. There's no getting around the paperwork and tax stamp, but you end up with a workable solution in weeks, rather than months. We covered the Form 1 E-File process in Issue 44, and the article is currently on RECOILweb.com, so if you haven't already, you may want to familiarize yourself with it. There's one departure from the procedure outline in the article, and that's the bit which deals with describing the manufacturer of the item.

On the drop-down menu, select FMI (for Form 1 manufactured), and you're on your way. You'll also have to describe the length and caliber of the can you're going to make. Tip: Some people get wrapped around the axle when it comes to fingerprints. There's absolutely no reason to make an appointment and pay a third party to fingerprint you, when you're perfectly capable of smearing ink on your own digits. Order a fingerprint kit from Amazon, and do it yourself in the comfort of your own home.

Once your Form 1 has been approved, which usually takes around three weeks, you can then buy a tube, spacers, baffles, and endcaps from the many online vendors that exist on the fringes of the interwebs. Due to the nature of NFA law, these will be described in rather coy terms, and you may wind up purchasing "barrel shrouds," "solvent traps," "oil filter kits," or "storage cups," all of which are largely useless for their advertised purpose, but give the vendors a fig leaf of deniability. Yes, it's all a bunch of bullsh*t, but it's the system we're stuck with.

Once your components arrive, you can then set to work engraving the tube to meet the legal requirements of the National Firearms Act (see RECOIL Issue 44). You could go get this done on a laser engraver and make it look all professional-like, or you could just bust out the Dremel. We did the latter, as it's going to be wrapped in a suppressor cover anyway. With your tube engraved, you can then drill holes in the baffles and endcap, screw everything together, and head to the range with your shiny new can. Enjoy! R

Our DIY suppressor consists of a grade 5 titanium tube and titanium end caps. The first two baffles in the stack are 17-4 stainless steel, and hardened with a propane torch, as these take the brunt of propellant gasses — the rest are titanium to keep weight to a minimum. Note the small cuts on the side of three baffles, designed to induce more turbulence and hence slow gas flow further; these were done with the ubiquitous Dremel and finished with a Swiss file.

SOURCE

ATF E-Forms: https://eforms.atf.gov/

Form 1 Suppressor forum: http://form1suppressor.boards.net/

Parts Vendors: https: //superprecisionconcepts.com/

https://sdtacticalarms.com/

A DIFFERENT KIND OF MODULAR SILENCER

Introducing the Novus From the Enfield Rifle Company

By Dave Merrill

A lot of engineers, entrepreneurs, and inventors all start the same way: exclaiming that something out there must be better.

And it was no different for Michael Tiziani. Years earlier while serving in Ranger regiment, Tiziani got his first taste of silencers; he fired suppressed Glocks, SCARs,

and an M110 sniper rifle or two. While he noted the benefits, he didn't like the perceived limitations of his issued equipment.

After he left active duty, Tiziani went to college for engineering and eventually became a full-time aerospace engineer. It was the combination of his background and education that ultimately became the impetus for the Enfield Rifle Company and their first mainstream product: the Novus silencer.

Before your mind fills with images of the godawful SA-80, this new American company doesn't have any relationship with the British rifles or the Royal Small Arms Factory (RSAF) that produced them. The roots of the owners land firmly in both Enfield, North Carolina, and Enfield, Connecticut.

When searching for a name, they found another Enfield Rifle Company; upon learning the owner was about to retire this Enfield Rifle Company, they bought the rights.

Please bear in mind that the example we have is a preproduction model; there'll be some changes prior to full production, and we'll note differences between the two as we go along.

A DIFFERENT KIND OF MODULAR

Right now when we think of modular silencers, images of removeable baffle designs such as the Dead Air Odessa or Torrent Suppressors Orthus spring to mind. While these silencers ship in the longest possible configurations, separate baffle sections can be removed or installed to customize length for the given application. That's not what the Novus is. Nor is it a silencer with different large module sections such as the SilencerCo Hybrid 36M. And while the Novus has replaceable endcaps, that's not how they do it either.

The Novus controls gas not with length, but with the apertures inside the silencer itself. Admittedly, when we first heard about this project we weren't sure how they'd go about it, but go about it they have.

Each Novus silencer can be configured with baffles rocking a .45, .30, or .224 aperture size — all without any spare parts that would make the BATFE give them side-eye.

Pistols or rifles — the Novus doesn't care so long as you have the right mount and caliber configuration.

DESIGN

The Novus doesn't represent Tiziani's first attempt — he went through several design iterations, including 3D printed and home-brewed design prototypes, before ERC began CNC operations. Nor is ERC a fly-by-night company using prefabbed parts from internet retailers as we've recently seen more of. Everything is made on-site; we've seen the lathes and CNC machines for ourselves in their Georgia production facility.

The body of the Novus is constructed from titanium and is 6.5 inches long, and the skeleton and baffles are cut from 17-4 stainless steel. All of this adds up to a weight under 15 ounces in a dedicated configuration and direct-thread mount.

As it ships from the factory, inside each Novus is a skeleton or core with 18 baffles nestled inside. You can shoot it in this configuration with .224 or below; but you'll probably never shoot it in this arrangement.

All you have to do to change baffle apertures is pop out all 18 baffles and choose six in the desired caliber. The skeleton is designed so that these Russian matryoshka baffles can be installed all together or six at a time. Each baffle has a clip that we're told should all be installed facing the same direction (more on this later).

Our preproduction model had an endcap with a .45 hole, but the company indicated that other end-caps such as caliber-specific ones or a flash hider/brake are likely to come in the future.

Instead of making their own proprietary mount, which is a market-killer for many a new silencer company, ERC went with the nearly universal 1 3/8-inch x 24 thread pattern designed by SilencerCo for their Omega suppressors. There are at least a dozen companies making adapters, mounts, and boosters with this pattern, giving you nearly

Here you see the progression of the Novus, from 3D printed concepts to hand-built prototypes to initial preproduction to what you see here today.

An engineer by trade, Tiziani has no problems going hands-on at the shop.

endless options. With that said, the Novus will ship with 5/8x24 and ½x28mm direct thread adapters included to get you going right away.

ENDLESS NERDERY

This suppressor isn't necessarily for the nerdy, but boy can you really, really play around with it.

Want to try out half 5.56 baffles and half .30-caliber baffles on a 5.56? There are endless combinations. Get really weird — want to install an upside-down .45 baffle as the first and then offset baffle clippings the rest of the way? You can do that too. If you like to tinker, then undoubtedly this is the silencer for you.

With many silencer designs, the blast baffle itself is specially reinforced because it takes the brunt of the damage. While that isn't present on the Novus, its longevity is incredibly high simply because you can rotate out your blast baffle whenever you want.

This level of modularity is the greatest strength of the Novus and will also be the subject of the most calls to customer service — namely, a consumer using the wrong baffles for the wrong firearm and causing strikes as a result. Always double-check your work before putting rounds downrange.

Our preproduction model doesn't have aperture size markings, so we had to rely on our eyeballs, but production baffles will have machined markings and will also be different colors (silver, gold, and black) to help prevent this sort of mishap.

AT THE RANGE

The Novus sounds exactly how we'd expect it to. A 5.56 rifle with a Novus in a 5.56 configuration would fit right in with its peers, as would a Novus in a 30-caliber setup. Even the 30-cal on a 5.56 sounded exactly as we'd expect. The performance of the Novus with a 9mm host and 45-caliber baffles was louder than most, except the smallest dedicated 9mm silencers, though we think a 9mm endcap (hint, hint ERC) could make a large difference here.

It's pretty safe to think of the Novus as a rifle silencer that can also play on a pistol rather than the other way around.

QUIRKS

Something a potential buyer will notice at the retail counter right away is that there's a rattle if you shake the Novus. That rattle is for a reason

Below: Our preproduction model wouldn't eat from every adapter — but production models will.

— there has to be enough clearance for you to take out the baffles to reconfigure them. This happens in both the 18-baffle and the caliber-specific six-baffle arrangement. However, the rattle quickly goes away with a little use. Five .300 BLK subsonics were enough to eliminate it the first time out.

We can tell you with certainty that not every 1.375-inch x 24 device or adapter will fit, for two reasons: Some devices may simply be too long, in which case the first baffle can be removed, or the mount itself may be too long, running into the base of the skeleton core, such as with our Gemtech bi-lock adapter built for a Dead Air Nomad.

Because there are so many companies making mounts and devices, you'll be bound to run into some compatibility issues.

However, this is one of those key differences between our preproduction prototype and the production model: The production model will be far more accommodating and will fit SilencerCo ASR and Dead Air Keymo mounts. Furthermore, you're unlikely to run into an issue with a muzzle device that's too long, because you can simply remove the first baffle if you need to.

LOOSE ROUNDS

We applaud Tiziani and the crew at ERC for designing a truly unique silencer. Normally, in order to achieve this level of customization the end user would have to be an NFA manufacturer themselves. While there are a lot of "do-all" silencers in the world, the ERC Novus is the only one out there with this kind of flexibility. **R**

ENFIELD RIFLE COMPANY

NOVUS

CALIBER:
.224 / .45 / .30

LENGTH:
6.5 inches

DIAMETER:
1.5 inches

WEIGHT (CALIBER-SPECIFIC, DIRECT THREAD MOUNT):
14.2 ounces

MSRP:
TBD, expected to be around $1,000

URL:
enfieldrifle company.com

CALMING THE BIG BORES

The Griffin Armament Bushwhacker 46

By Dave Merrill

As more and more states allow for straight-walled rifle rounds for hunting in traditionally shotgun-only areas, we've seen an increasing interest in big-bore loads such as the .450 Bushmaster and, more recently, the .350 Legend. Calibers like this pack a punch both in terms of recoil, as well as sound and blast.

It was only natural that we'd start seeing more silencers fit for this role, and the latest from Griffin Armament is exactly this. The Bushwhacker 46 plays well with everything mentioned above; it also transforms into a pistol suppressor if needed. What you see on these pages is a production, not a preproduction model.

DESIGN

The 8.25-inch-long tube is turned from 17-4 stainless steel with a black nitride finish, and the baffles are made from 6AL4V titanium. The finish on our example was deep and uniform, which we've come to expect from Griffin

GRIFFIN ARMAMENT

BUSHWHACKER 46

LENGTH:
8.2 inches

WEIGHT:
16.3 ounces (rifle configuration), 18.1 ounces (pistol configuration)

MSRP:
$999

URL:
www.griffinarmament.com

The Bushwhacker handled 350 Legend like it was made for it, which it sort of was.

Armament. The stainless endcap is removable and interchangeable with Griffin's Paladin line of silencers. It may seem a bit long for some, but keep in mind that this is designed for the likes of .450 Bushmaster.

With increasing modularity on their minds, Griffin Armament jumped on the SilencerCo Omega 1.375x24 threaded-rear-end bandwagon, which means there are mounts and options galore. The Bushwhacker itself ships with a Griffin Plan-A 1.375 mount included, along with an accompanying 5/8x24 Minimalist 30-caliber brake, as well as a new compatible pistol booster. The Bushwhacker doesn't come with the piston required for the pistol booster, so you'll have to purchase that one separately if you want to use the Bushwhacker on a pistol. However, if you already have an appropriate piston by Griffin, SilencerCo, or Rugged, you'll be all set, as you also will be if you have a Dead Air piston designed for the Odessa (no-go for the rest of the Dead Air pistons).

AT THE RANGE

Our foremost thought was to mount the Bushwhacker 46 on a .350 Legend for a hunting rig. While the silencer is over-bored for that caliber, it almost meant that it kept the gas pressure down on our rig. While OSHA wouldn't like the sound level, it was a helluva lot better than a simple flash hider on the end. Undoubtedly there would be a few more decibels taken off the high end if we used a Paladin endcap specifically made for .350 Legend (which doesn't currently exist — hint, hint, Griffin).

There hasn't been a complete set of testing data made available, and even then there are arguments to be made about how testing should occur. With that said, the initial numbers we were provided matched our subjective impressions at the range — 137dB from a 16-inch 7.62N, 130dB dry from .45ACP pistol, and the loudness level pushing into the upper 130s with a 16-inch 5.56mm.

LOOSE ROUNDS

Like many general-purpose silencers, this one did "alright" as a multi-caliber silencer, but that's really not what this can is for. This is a big-bore suppressor that also plays well with others. If you're looking for a general-purpose silencer, this one may not be for you, but if you have a use-case for suppressing .450 Bushmaster or .458 SOCOM, you'll no longer be stuck with something much larger originally designed for calibers like a .50BMG. **R**

PURPOSE-BUILT FOR PRECISION

Area 419's Maverick Silencer Gets Goosed for Full Production

By Dave Merrill

Chances are that if you're even casually interested in precision shooting, the name Area 419 won't be new to you. At their Ohio facility, they make all manner of muzzle devices, adapters, reloading tools, and a whole lot of parts and pieces that go in and around precision rifles. And now, silencers. This last November, we were the first to show you some preproduction models of their new Maverick suppressor. As is the nature of many first iterations, there were some changes in the following months. What you see on these pages is their full production model, quieter and more modular than before.

DESIGN

Originally, the Maverick was a two-piece design, allowing the user to choose between a small suppressor with a large, effective brake attached and a longer more traditional silencer configuration — and you can still do that. What the folks at Area 419 did with their full production model was offer even more options to outfit the Maverick to your intended role.

Phase 1 is your base. It features a larger, .45-caliber bore and unclipped baffles. A brake or endcap can be fitted to the end of Phase 1 either to make a very short linear compensator or to reduce the sound signature when a brake is in use.

Phase 2 is caliber-specific, with .30 caliber and 6.5mm as initial offerings. This is where you'll have some significant reduction in sound — but they're not done yet.

While this was the original preproduction design, Area 419 thought they could eke out some more performance with a minor update: the Phase 3. This little modular .45-caliber baffle can be added to any stage to further reduce sound, while adding only a modest amount of length.

All told, there are six different configurations for the Maverick. Want something that just takes the edge off? You can roll with a mere 4-inch base module. If you want something less obnoxious, toss the Maverick brake on top.

Aside from the addition of the smallest phase, the engineers at Area 419 made some changes to the baffles themselves, not only quieting the entire unit down but also ensuring you won't see any signifi-

cant point-of-impact shift regardless of how you run the Maverick.

As expected, there are no low-end materials here, as the entire body is turned from a solid piece of titanium. The brake, mount, and blast baffle are made from stainless steel. In the absolute heaviest possible configuration, the entire Maverick weighs in at less than a pound. The lightest? Just over 7 ounces.

We're told that eight configurations will be available for purchase: Your choice of a Hellfire or Sidewinder mounting system, either a 30-caliber or 6.5mm Phase 2, and with or without the additional small module. There's around a 9-percent cost increase when you add the small module, but the flexibility it provides is worth it.

AT THE RANGE

Keep in mind that the Maverick (and indeed all Area 419 mounts) are threaded left-hand; therefore, you won't inadvertently unscrew a muzzle device when removing the silencer or adapter. We strongly recommend the use of anti-seize on the threads between modules, lest they seize up.

We weren't dissatisfied with the original Maverick silencer, but we did see some room for growth. With the first pull of the trigger on a 6.5 Creedmoor rifle, we immediately noticed some sound reduction relative to the preproduction version. Area 419 claims the 9-inch setup meters 2 decibels less than the SilencerCo Omega. But it's not all about the sound signature or the meter.

Though we can't say there's absolutely no shift, we can tell you that, within the realm of practical accuracy for PRS shooting, we couldn't detect any differences when going the distance. But bear in mind, this is a precision silencer built for precision rifles. The barrels themselves are invariably thick and stiff, and the rifles they're mounted on are stout all by themselves. If we hung a Maverick off of the end of a 2-inch pencil-barreled AR-15, it might be a different story, but that's not what this silencer is designed for.

LOOSE ROUNDS

Anything that makes those godawful brakes precision shooters like to use less godawful is a welcome addition, and the fact it's paired with an effective silencer is extra icing on the cake. Long range shooters are always looking for consistency, and Area 419 promises exactly that with the Maverick, as well as reducing sound. We'll continue testing that statement; stay tuned. R

You can switch from a brake to any Maverick configuration, without tools, right at the range.

Options? You've got options.

AREA 419
MAVERICK

LENGTH:
4 inches (shortest) to 9 inches (longest)

WEIGHT:
7.3 ounces (shortest) to 15.2 ounces (longest)

MSRP:
$1,300 for the 2-piece configuration; $1,500 for the 3-piece

URL:
www.area419.com

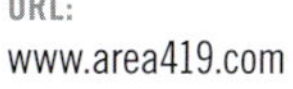

TOOLS OF THE TRADE

Like an LS-swapped Volvo, Our Editor's Personal Carbine Looks Homely, But Has Performance Parts Where it Counts

By Iain Harrison
Photos by Kenda Lenseigne

This is my work gun. There's nothing outwardly fancy about it. No custom Cerakote, no go-fast lightening cuts, and its grimy, carbon-crusted, battered exterior won't gain many likes on the 'Gram, but it's been through numerous training courses, visited the Tactical Games, survived InRange TV's Desert Brutality match, and is the go-to carbine for anything that needs a hole putting in it. In short: It just f@cking works.

All its components have been selected based on personal experience, which is to say, your mileage may vary, and you probably have just as strong preferences. But there's something to be said for having access to just about any high-speed parts from the vast panoply available for this 60-year-old design, which has been through more revisions than a Kardashian's face.

At its heart is a 12.5-inch barrel from HM Defense, the length chosen due to hitting the sweet spot between external ballistics and maneuverability. It's a monoblock design, meaning the gas block and barrel are carved out of one piece of 4150 CrMo steel, which eliminates three components and a potential failure point. There's a reason Knight's Armament thread their gas blocks onto the barrel, and this takes things one step further. The

bore is triple lapped and button rifled prior to nitriding, and is a midweight profile that holds its group size fairly well under lengthy strings of fire. Its mid-length gas system makes for a remarkably soft-shooting SBR, and it handles both suppressed and unsuppressed duties with ease.

Its bolt and carrier are from Extreme Defense and are about as far as you can take the lessons learned about the DI operating system while still remaining within the realms of earthly material science. The bolt itself features revised geometry of both locking lugs and cam path, in order to increase strength and stave off the usual causes of failure — fracture of the lugs adjacent to the extractor and breakage at the cam pin hole. The longer cam path also serves to delay unlocking, giving a smoother operating cycle and reducing stress on components. One of the notable revisions to the design of AR-10-type carbines, such as those fielded by Daniel Defense and the now-defunct DPMS, was the adoption of twin ejectors, which provide redundancy and more positive clearance of spent cases. The Extreme Defense bolt manages to shoehorn a second ejector into a 5.56 bolt face, and provides the same advantages. Finally, everything is finished in an incredibly slick coating, which is light-years ahead of the nickel boron BCG's we've run in past versions.

Up front, there's a SIG flash hider, which accepts a SRD556 can, and its SLR Rifleworks 11.7-inch ION rail is sized to accept virtually any other suppressor without interference, should something better come along. The rail provides mounting points for the latest SureFire Scout Light and Cloud Defensive LCS tape switch housing, which in combination take care of illumination tasks. At the 6 o'clock position lurks a mount for a Spartan Precision QD bipod, which rides on the shooter's belt when not needed, but provides a major advantage when it comes to making hits at extended ranges.

The carbon fiber and aluminum bipod allow the shooter to fully exploit the capabilities of the Nightforce NX8 1-8x24 scope. At just 18 ounces, the NX8 is tough as nails and light enough not to be out of place on an SBR. It clamps to a BCM Mk2 upper, which according to the manufacturer is 30-percent stiffer than a regular M4, for a weight gain of 1/3 ounce. Stiffening the upper means there's less deflection on the barrel nut from rail-mounted accessories, and both upper and Radian Raptor SD charging handle work together to limit the escape of gases back toward the shooter's face.

The part of the gun that's on the NFA registry is made by American Defense Manufacturing and was chosen both for its quality and ambi controls. One of only three makers who can claim to be fully ambidextrous, including bolt hold open and release, ADM UIC lowers are intuitive and, should you not care for the ability to run the gun exactly the same way whether it's in your right shoulder or left, can be operated like a regular AR-15. **R**

PARTS	MSRP
Lower: ADM UIC	$600
Upper: BCM Mod 2	$125
Barrel: HM Defense 12.5-inch Monobloc	$300
BCG: Extreme Defense	$285
Trigger: Geissele SSA	$240
Charging Handle: Radian Raptor SD SL	$105
Rail: SLR Rifleworks ION Lite 11.7-inch	$260
Primary Optic: Nightforce NX8 Secondary Optic: Leupold Deltapoint – superseded by the DP Pro	$1,750
Mount: Aero Precision Ultralight	$110
Muzzle Device: SIG SRD	$71
Suppressor: SIG SRD 556 QD	$750
Sling: Lead Faucet Tactical	$55
Bipod: Spartan Pro Hunt	$330
Total	**$4,981**

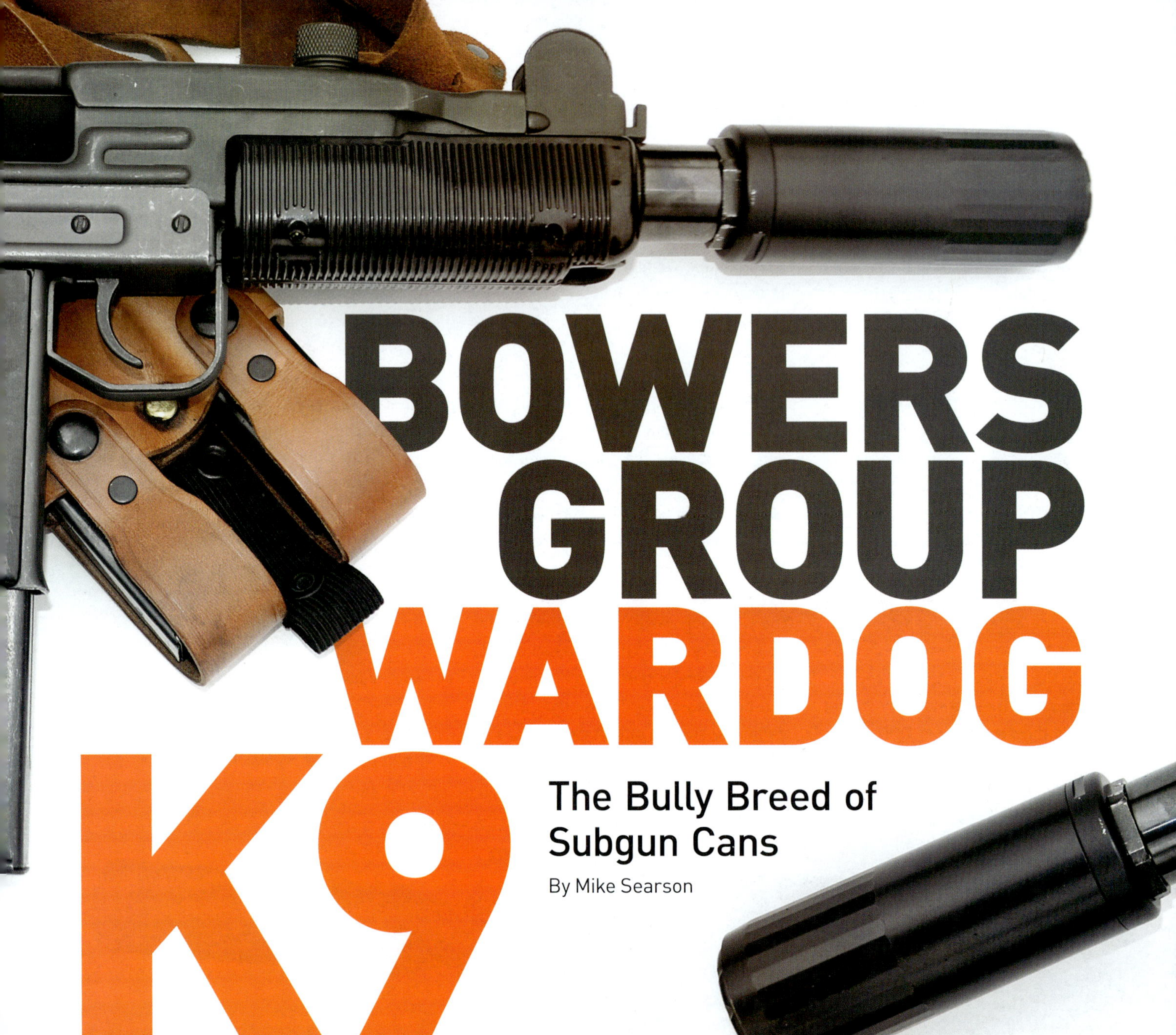

BOWERS GROUP WARDOG K9

The Bully Breed of Subgun Cans

By Mike Searson

Shorter, lighter, louder. That's how Tom Bowers describes his latest silencer, the Wardog. Coming in at 1.7 inches in diameter by 4.3 inches in length, it tips the scales at a mere 7.8 ounces. Bowers claims it's rated for heavy use in full-auto fire with both 9mm and 300 Blackout, subsonic and supersonic. One of the ways they were able to keep it so svelte was by using aluminum for most of the construction with steel inserts, which will absolutely be a turn-off for some. But read on and remember: Short silencers are the definition of compromise.

The Bowers Wardog is the shortest 9mm subgun can on the market — so short it can be mounted to the gun and carried in a DeSantis Secret Service shoulder holster.

DEVELOPMENT

The Wardog K9 has its roots in the Bowers Group's Vers 9, which is a full-sized, high-volume 9mm subgun silencer rated for full-auto fire. Because it's in 9mm, it can perform double duty as a host for subsonic 300 Blackout.

There was one problem with the Vers 9 — its size. It has an overall length of 11.25 inches and weighs a good 18 ounces. Not too bad for a sub-gun can with five baffles, but kind of ridiculous to hang on the end of a lever gun or bolt gun with a 16-inch barrel.

So, Tom went to the drawing board and determined that he could remove 2.6 inches, a baffle, and 3 ounces of weight. The end result was the Vers 9S. Now it's arguably one of the best submachine gun silencers ever made, but Tom hit a snag with this design a few years

ago: He bought a Ford Raptor and found that his Uzi SMG would fit in the door pouch, but not with a Vers 9S attached.

Once again, Tom changed his design. In order to get to the required length, he removed two baffles from the Vers 9S and took it down 2.2 inches to 6.4 inches in length. With the right 147-grain subsonic ammunition, the can meters at 138 dB. It fit perfectly in his ride, but he was still reluctant to release the design to the general public, because it was only barely hearing safe.

UZI DOES IT

There are two ways to approach suppressing an Uzi submachine gun. You can go the threaded barrel route or use a mount, which replaces the barrel nut. We find the barrel nut to be superior, as it doesn't tend to loosen up from the torque and heat generated by full-auto fire. If you're running a semi-auto Uzi, you can probably get away with a threaded barrel. But for full-auto fire? Give us the nut.

Even with mouse fart subsonic ammo, the Wardog was barely hearing safe. It wasn't uncomfortable but wasn't close to "Hollywood Quiet." With 115- and 124-grain ammunition, you'll want to wear ear-pro, but that's the price you pay with virtually any Kurz can (see RECOIL Issue 36 for more on Kurz cans).

Our first range outing with the Wardog K9 and the Uzi resulted

Although designed for a 9mm sub gun, we found it more effective on a Marlin 1894 lever action chambered in 357 Magnum and 38 Special.

in numerous failures to fire. Light primer strikes were the culprit. We ripped off the top cover and examined the bolt and firing pin. They were all within spec, as was the ammo.

It turned out that it had to do with the Uzi adapter. The fit on these isn't always perfect, and sometimes there's a gap, which allows the barrel to move back and forth. This is solved by using wave washers to prevent barrel movement. Our washers completely flattened, and we solved the problem by adding two wave washers to decrease the gap. It goes to show: Always check first. Otherwise, you could risk an out-of-battery malfunction that could result in injury or permanent damage to your Uzi.

LEVER TIME

We then turned to the one platform we were hoping the Bowers Wardog K9 would excel: crowning a Marlin Model 94 CST chambered in 357 Magnum. Elsewhere in this issue, we have an article about another Bowers silencer mounted on a different lever-action rifle. This isn't the latest rifle from Marlin to sport barrel threads. Ours is about 10 years old and chambered in 357 Magnum. It boasts no rails or scope mounts but wears a Skinner rear peep sight and Bear Buster front sight.

Swapping out the Uzi insert for a Versadapt insert in ½x28, we were ready to go.

On the Uzi, the sights sit high enough above the barrel that the suppressor isn't an issue; on a skinny lever gun like the Marlin 1894, the target is going to get covered. You can get around this by using an optic or by "shooting through the can."

In 38 Special, the rifle sounded extremely nice with 158-grain lead round-nose bullets that were probably cruising at 1,000 feet per second from the 16-inch barrel. A jacketed hollow-point load of the same weight in 357 Magnum wasn't as pleasant. That projectile had to be approaching 1,750 to 1,800 feet per second, and it was quite loud.

In our estimation, the light weight of the Wardog K9 and its short size make it fine for a lever-action rifle. The closed and completely sealed action with a longer barrel makes for a superior host compared to an open-bolt short-barreled SMG.

KALASHNIKOV USA KP-9

Even though it's also blowback, the AK-9 was noticeably quieter than the likes of the full-auto Uzi with its open bolt. Generally, blowback is louder than other actions suppressed because too much gas escapes the ejection port

BOWERS
WARDOG

CALIBER
9mm

CONSTRUCTION
Aluminum with steel threaded inserts

LENGTH
4.3 inches

DIAMETER
1.7 inches

WEIGHT
7.8 ounces

MSRP:
$635

URL:
bowersgroup.com

during cycling, but an open-bolt is even worse.

For fun and science, we also ran it with a Franklin Armory BFS III binary trigger. While a gas-operated PCC would sound better, the AK-9 and Wardog combination is a good, compact, and decently effective setup. Again, it's not the quietest, but it's short, keeps the overall weight down, and is right at the threshold for being hearing safe (but our ears aren't OSHA rated).

LOOSE ROUNDS

Given our druthers, we'd select a silencer made of a more robust material than aluminum for select-fire, but every aspect of any Kurz can is a search to find middle ground. The Bowers Wardog is a versatile suppressor if you have hosts in 300 Blackout, 9mm, 357 Magnum, and 38 Special. With the right host, it can make a quiet, lightweight hunting can that's not going to add an extra 9 or 12 inches to your rifle's overall length.

Other hosts, like the Uzi and AK-9 will barely make it quiet without the correct type of ammunition. If you're running a 16-inch PCC in 9mm, you'll achieve better results simply due to the longer barrel — and it doesn't stick out too far either.

This is definitely not a "first can" for most shooters. Rather, the Bowers Wardog is a niche can for shooters who don't necessarily want the quietest but do want something shorter and lighter and don't mind it a bit louder.

Or if you're looking to stash your suppressed subgun in your truck's door pouch. **R**

MODERN-DAY DE LISLE CARBINE

THE CURTIS TACTICAL CT700P INTEGRAL 9MM

BY DAVE MERRILL

It started on a whim while in a rural town in Ohio looking for a local gun shop to peruse for a little bit of business and mostly grins. The Google map highlighted "Curtis Tactical Suppressors," and the journey began. While driving down a dusty country road, the realization hit that this may either be an incorrect address or a small backyard shop, but plenty of damned good work takes places in garages and sheds across the great American nation. There was no hesitation.

Mowing the lawn in front of his house, the location of his former backyard shop, was Joe Jones. Jones is the founder and owner of Curtis Tactical, and by mere happenstance he happened to take a day off to perform yardwork on his large property. Introductions were made, and soon we were firing machine guns behind his house and learning details of his wares.

It was totally random and awesome. We blame a glitch in the Matrix. Or Elon Musk. Same/same.

Though Jones has been designing and building silencers for nearly a decade, it wasn't until mid-2019 Curtis Tactical was officially formed. Curtis Tactical made a name for themselves by re-coring and otherwise upgrading silencers of all types (if you have a now-defunct Huntertown Arms silencer, give them a ring). Jones pulled nearly endless examples of upgraded suppressors from his safe from names both big and

small. Like us, the more purpose-driven the suppressor, the more he liked them. And nothing is more purpose-driven than an integrally suppressed weapon (see RECOIL Issue 44 for a full breakdown on integrals). He had modernized interpretations of older guns: Integral 9mm MPXs replaced the vaunted MP5SD. Nazi-killing OSS High Standard HDMs were present but with 2020 baffle designs. And what really caught our eye? A rifle that got weirder and more astonishing the more we learned about it: the CT700P.

The CT700P is a Remington 700, integrally suppressed, chambered in 9mm, and converted to controlled feed, that eats ammunition from Glock mags.

The CT700P immediately brought forth images of the De Lisle. The De Lisle carbine (sometimes called the De Lisle Commando) was bizarre British invention, which was essentially a Lee-Enfield MkIII converted to .45 ACP using a Thompson submachinegun barrel and integrally suppressed — furthermore, it ate from modified 1911 magazines. The De Lisle saw use among British commandos and SOE for the purposes of silent Nazi-zapping. And let's be real: Any gun used to covertly kill Nazis will always be among the coolest guns on earth.

A rifle we didn't even know we wanted, which we instantly knew we'd have to get our hot little hands on.

Keep in mind that the rifle you see on these pages is a preproduction model, and we'll note potential changes to full production models as we come across them.

ACTION AND OPERATION

Bolt actions make for the quietest possible silencer hosts.

It looks like a normal bolt gun — until you get to the magazine.

Ejection port noise is nonexistent, since the bolt is opened only well after the projectile and all gases leave the barrel. The CT700P is a single-stamper, because the barrel is technically 20 inches long with an 8-inch silencer popped on the end of the 12-inch rifled portion.

Each CT700P rocks a receiver machined from 6AL-4V titanium and is DLC (Diamond-Like Carbon) coated. The bolts themselves are machined from 17-4 stainless bar stock, feature threaded handles if you want to swap, and are UltraOX nitride.

The briefest explanation possible of controlled feed versus push feed is that a push-feed action has an extractor that snaps over the rim of a cartridge only when fully chambered (like an AR-15), whereas a controlled-feed mechanism has control of the cartridge with the extractor through the entire cycle of operation like a Mauser model 98. Jones contends that there's no other way to

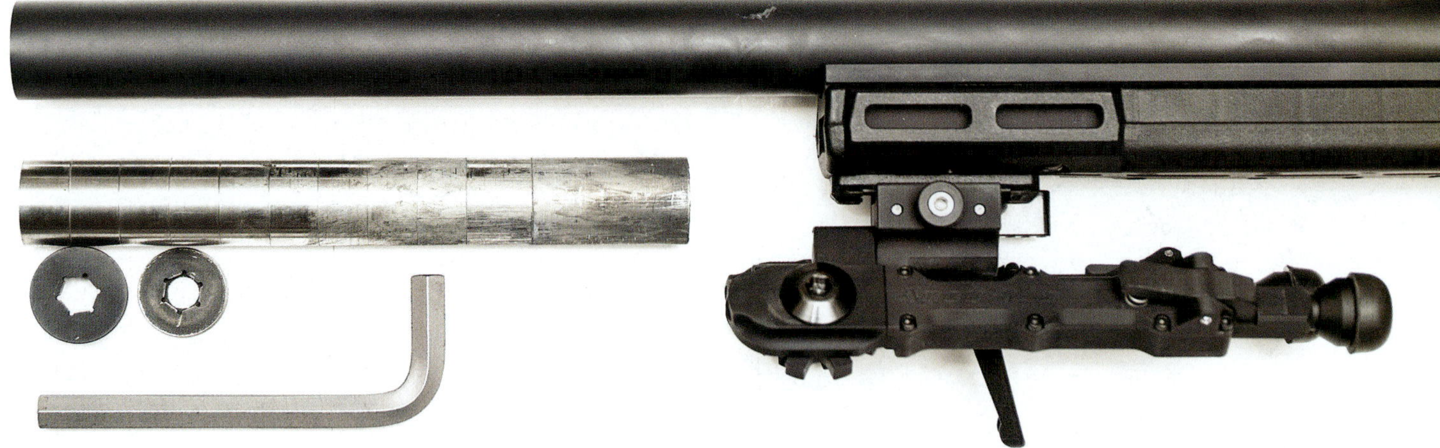

feed a pistol round with a front-lug bolt action than a controlled feed system, and he's also spent an awful lot of time developing this system expressly for that purpose.

Jones further explains, "The bolt face is where design is critical to get the pistol rounds to feed in a front lug action. Our patent-pending angled T-slot machined into the face of the bolt makes this a true controlled round feed and is necessary to feed the pistol cases. As the round exits the magazine, the T-slot on the front of the bolt captures the case rim; it's machined to the perfect profile to allow a smooth feed up the feed ramp into the chamber while being retained the whole time. The side-mounted bolt stop also plays a big role in the ejection; as the bolt is moved rearward the bolt stop actuates the ejector, which is spring loaded to the rear."

The first CT700P we shot required a rather aggressive operation of the bolt in order to fully seat each round, but this problem nearly disappeared on our example due to a slight extractor tweak and shouldn't be present on production models.

Instead of a modified Accuracy International magazine (see page 104 for an example of a sub-caliber magazine conversion) inserted into the KRG chassis there's an adapter for Glock magazines. Our preproduction sample can seat Glock 19 magazines and longer with ease, but the diminutive Glock 26 mags are too short to fully seat.

The baffles are held in place with both an endcap and a locking nut.

The extremely short action threw us off the first time we shot it.

The silencer portion of the CT700P is easily removed for maintenance with the included 7/16-inch hex key. First, the endcap is unscrewed followed by a secondary locking nut with the same hex key. After that, the snap-together core slides out. The tube remains permanently attached to the rifled portion of the barrel, giving us that legal length of beyond 16 inches.

Each baffle features symmetrical clips and each subsequent baffle should be 90 degrees offset from the previous for best consistency.

QUIRKS & WEIRDNESS

The magazine adapter itself can be removed like any other detachable magazine, necessitating the removal of the enlarged KRG release lest we inadvertently pop it out on the range. A new lever release on the physical adapter itself allows for the 9mm Glock magazines to be removed and inserted. At least with our preproduction, it's best to load a filled magazine on an open bolt to ensure proper seating. Along similar lines, the current adapter allows the magazine to over penetrate slightly, so after it clicks in, give it a little tug downward.

VARIANTS

Though you're seeing an integral 9mm on these pages, Curtis Tactical has several pistol-caliber models available for purchase. All models ship with 20-MOA scope bases, medium-contour threaded barrels, Trigger Tech triggers, the appropriate magwell adapter, and a magazine.

You can get a CT700P chambered in .40 S&W that uses small-frame Glock magazines, and also .45 ACP with the choice of either G36 mags (weird) or full-on De Lisle with 1911 mags. Rifle configurations have either 16- or 20-inch barrels and custom "Pork Swords" pistols (see RECOIL 50 for an example) with barrel lengths between 3 and 12 inches.

While any Remington 700 short-action stock or chassis can be used, standard options are KRG (shown) and Masterpiece Arms. This preproduction CT700P was put together with a Green Mountain 1/10 twist barrel; Douglas and Krieger precision barrels will be available with an upcharge.

MAD SCIENTIST

Seeing as how the 350 Legend uses the same projectile diameter of a standard 9mm, RECOIL Editor-in-Chief Iain Harrison loaded up some Hornady 170-grain FTX and Spire Point 350 projectiles in 9mm cases using 3.0 grains of WSF powder. While the results won't directly feed from Glock magazines, they will from modified 1911 mags — and it just so happens that Curtis Tactical makes one. We don't have full load development, yet these are quieter than pissing in the snow. Expect to see more of our 355 Whisper in the near future.

The angled T-slot allows for a true controlled round feed.

OUTFITTING

While this is a 9mm rifle, it's also a precision rifle. Extreme Long Range shooters have been known to push calibers traditionally considered short-range to the distance (1K yard shots with .22LR is increasingly more common these days among the ELR crowd) — we didn't push the capability envelope too far. Magnified optics were a no-brainer, but we decided an LPVO would make for the best-case use for our imaginary Nazi-snuffing scenario. After much studying of ballistic calculators, we settled on a 1-6x24mm Vortex Viper PST in a set of Precision Reflex Inc rings for the optic.

While we knew that standard Glock magazines would be fine, we threw the original game plan entirely out the window and threw a generally garbage stolen KCI 50-round drum magazine into the mix. No, these magazines aren't recommended for semi or automatic use, but they were perfectly fine for the slower bolt-action CT700P, though they certainly added onto the weight.

The KRG Bravo chassis allows for easy attachment of a sling and bipod, so we rolled with an Accu-Tac bipod and a legacy-padded QD Ares Armor Huskey sling.

AT THE RANGE

Shooting 9mm not from a pistol or even a carbine changes a lot of metrics. As you extend your range, you'll quickly find that the ballistic coefficient of your load matters far more with the CT700P than your pocket pistol. Our suggestion is to take several different types of ammunition to the range and see what she likes.

Using a MagnetoSpeed and an Applied Ballistics calculator, we determined our best zero using Israeli 158-grain ammunition

The baffles can be snapped apart for maintenance.

would be at 15 yards. This gave us pointy-clicky to 120 yards and easy holds for an additional 50 yards beyond.

For testing accuracy, we replaced the lower-power 1-6x Vortex with a 6.5-20x Leupold VX-3I LRP to get our eyeballs closer to the target.

Using IWI 158-grain FMJ 9mm, the average five-shot group was 1.15 inches at 100 yards. Curtis Tactical tells us their Kreiger and Douglas barrels are sub-MOA, provided you're using match 9mm.

Swapping grain weights of ammunition became troublesome, because cheap 115-grain Walmart Winchester white box has significant ballistic deviation at range when compared to their heavier (and quieter!) subsonic brethren. Our take is that the best practice is to stick to a single grain weight and brand of ammunition lest you either have to constantly re-zero or guess at projectile drops at range.

Given the longer barrel, baffle stack, and bolt-action, it should come as no surprise that the CT700P is less than handclap-quiet when loaded up with subsonics.

LOOSE ROUNDS

Coming across Curtis Tactical through random happenstance and being extroverted enough to walk up to a random house was the first of many new surprises. Sometimes backyard shops grow into sustainable businesses, and that nearly always happens when someone brings a new idea to fruition or takes an old idea and puts a new spin on it.

Sure, there are a few things that have to be worked out, but that's the nature of R&D for preproduction guns. Just while writing this article, the bolt and extractor design was tweaked twice — Jones is quick to make changes, and shifting gears quickly is something a small, custom shop can do.

Some would argue that the CT700P should accept Sig Sauer M17/18 magazines to be a true successor to the De Lisle, and to those folks we'd point out the bevvy of 9mm Glocks currently in USSOCOM inventory. We don't know what Curtis Tactical will cook up next, but we're for damned sure watching and waiting. R

CURTIS TACTICAL INTEGRAL CT700P

CARTRIDGE: 9MM LUGER

OVERALL LENGTH: 39 INCHES

WEIGHT UNLOADED: 9 POUNDS, 6 OUNCES

BARREL LENGTH: 20 INCHES

MSRP: $2,075

MAGAZINE CAPACITY: 10, 15, 17

URL: CURTISTACTICAL SUPPRESSORSANDRIFLES.COM

FEATURED ACCESSORIES:
JVORTEX 1-6X24MM PST VMR-2 $900
PRECISION REFLEX INC RINGS $5,590
SPUHR SP-5602 SCOPE MOUNT $108
ACCU-TAC BR-4 G2 BIPOD $299

PRICE AS FEATURED: $3,382

RECOIL

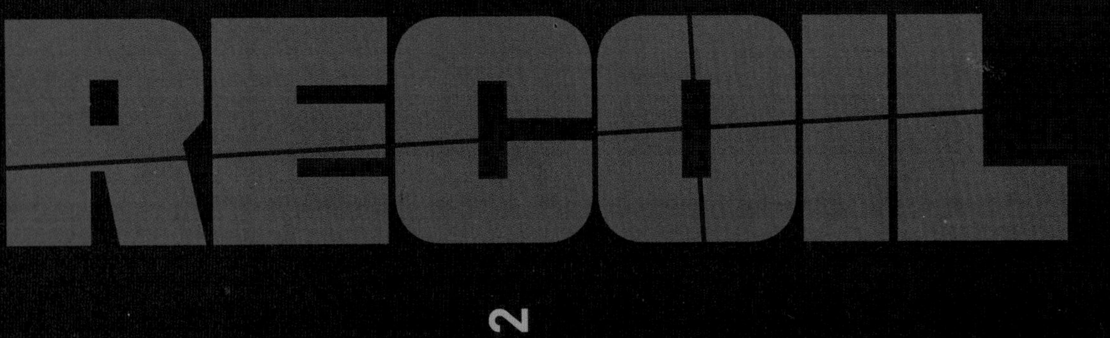

Dead Air Odessa

Liberty Sovereign

SureFire SOCOM556-RC2

Dead Air Nomad

Gemtech GMT-Halo

Gemtech Halo

SilencerCo Chimera

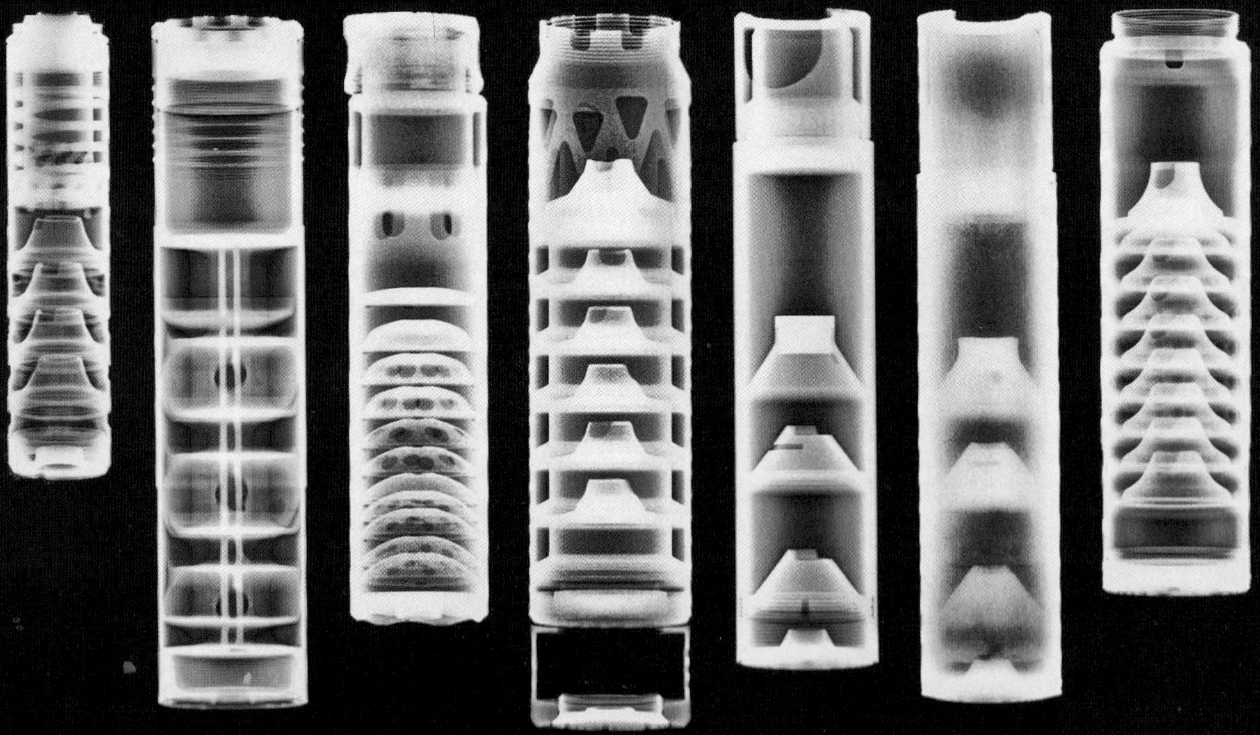

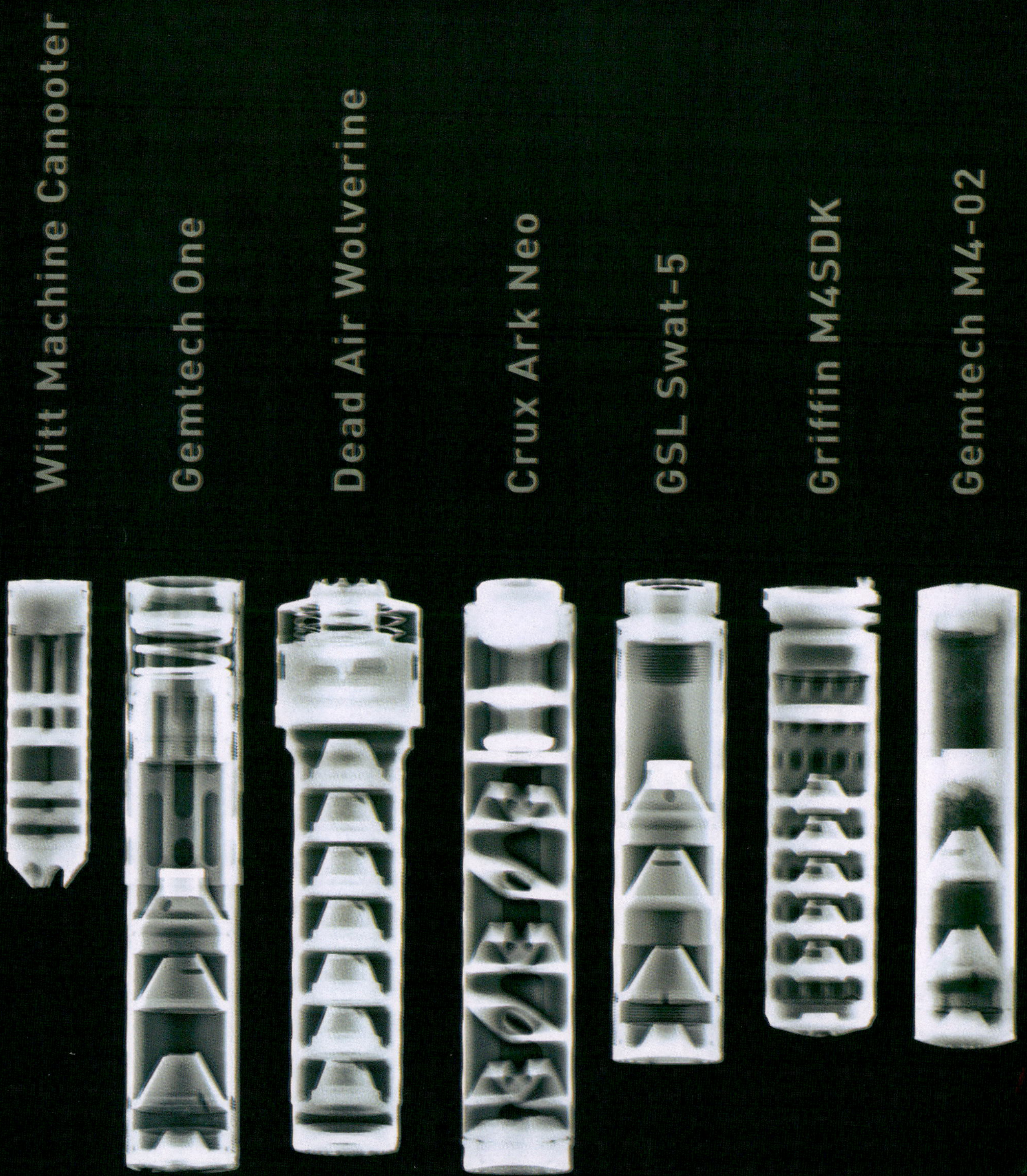

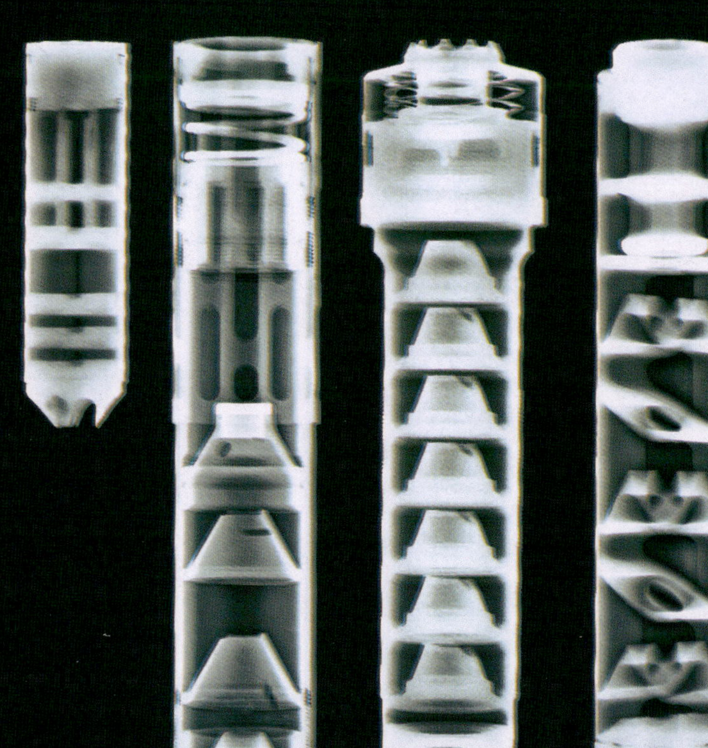
Witt Machine Canooter
Gemtech One
Dead Air Wolverine
Crux Ark Neo
GSL Swat-5
Griffin M4SDK
Gemtech M4-02

AAC Ti-Rant 9
Dead Air Wolf-9SD
Dead Air Sandman-L
Dead Air Sandman-S
Q LLC Half Nelson
TBAC Ultra 9
TBAC Ultra 7

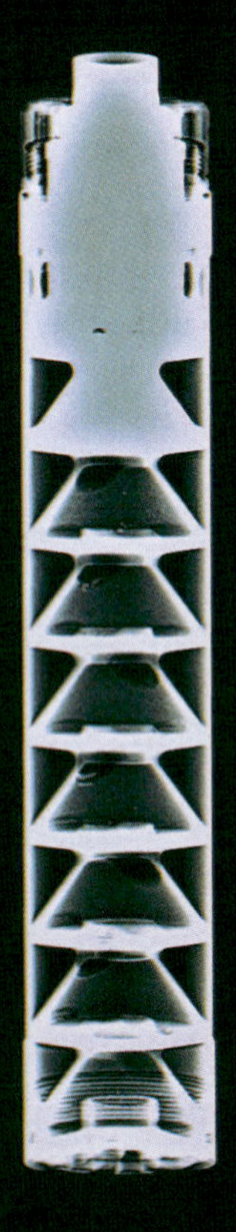

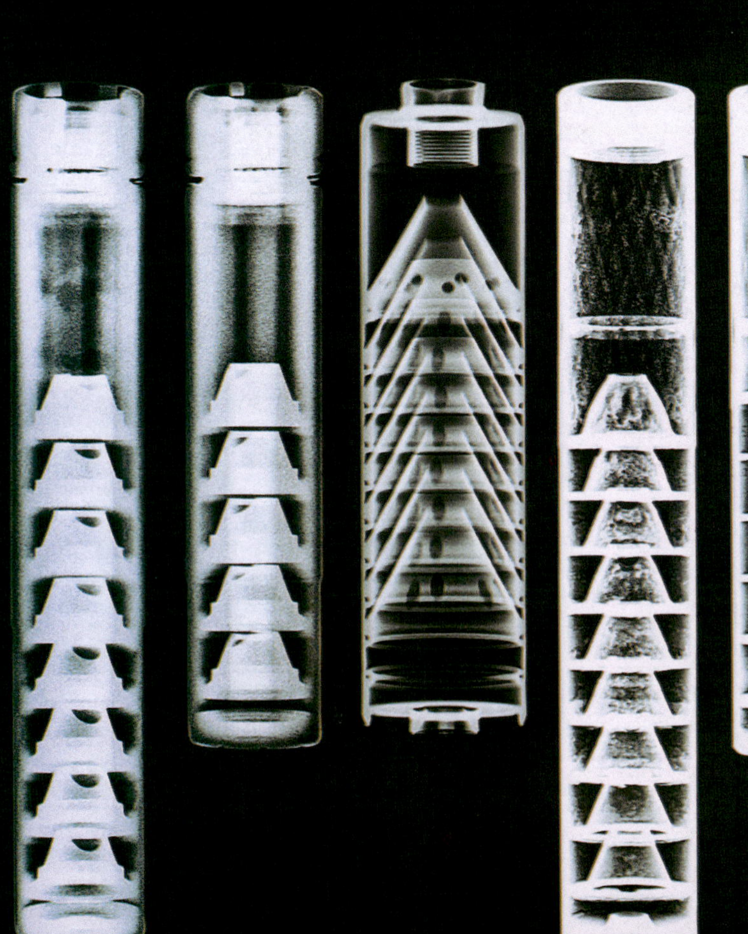

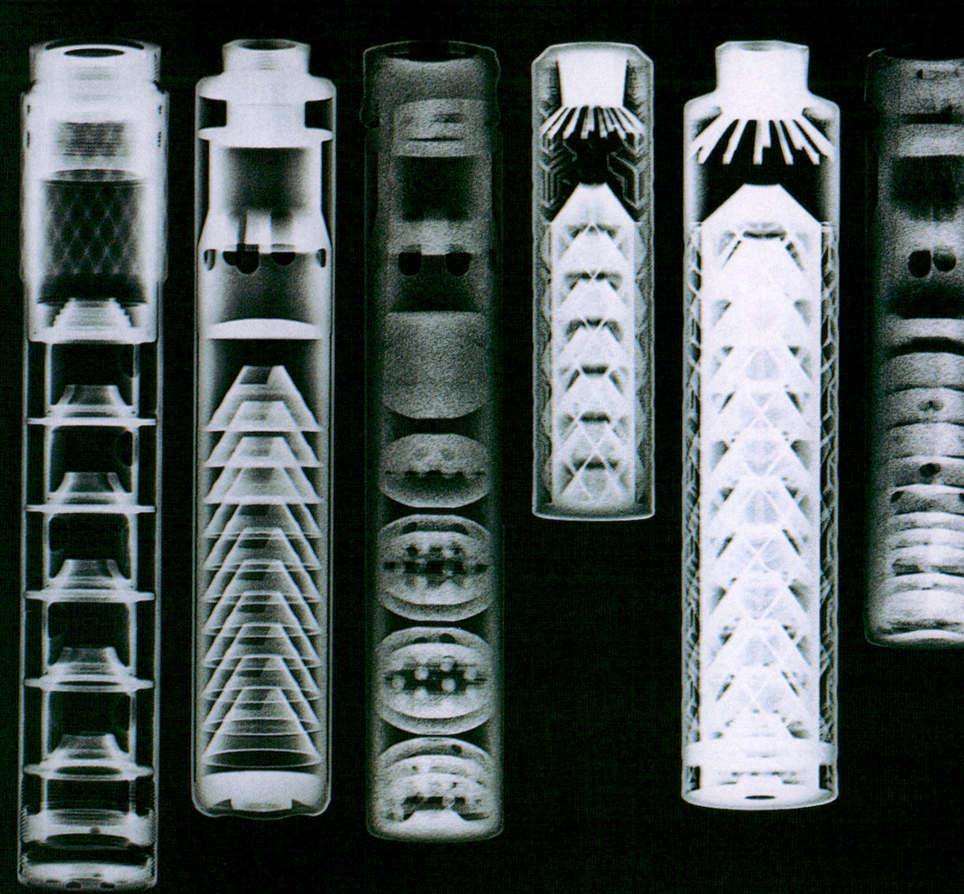
Silent Legion SL-MAG
SureFire Genesis 762
SureFire SOCOM762-RC2
TDS Bantam II
TDS STRIX II
SureFire FA556-212

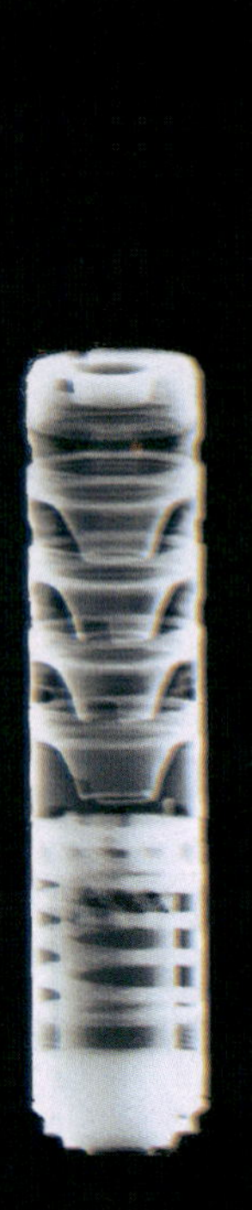

Dead Air Odessa

Make: Dead Air
Model: Odessa-9
Weight: 8.75 ounces
Length: 8.5 inches and shorter
Diameter: 1.1 inches
Caliber: 9mm
URL: deadairsilencers.com

The Dead Air Odessa has numbered baffles for correct assembly, which is also handy for knowing which baffle is still rolling around the back of your safe.

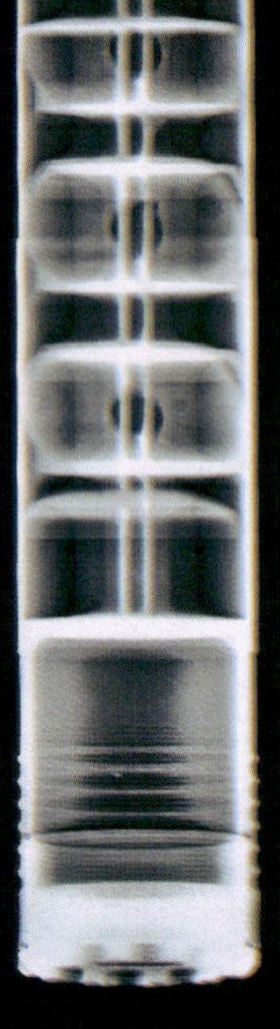

Liberty Sovereign

Make: Liberty Suppressors
Model: Sovereign
Weight: 12.7 ounces
Length: 7.125 inches
Diameter: 1.625 inches
Caliber: .30
URL: libertycans.net

The Sovereign was originally designed for hunters using bolt-actions but often finds a home on one of our 7.62N ARs.

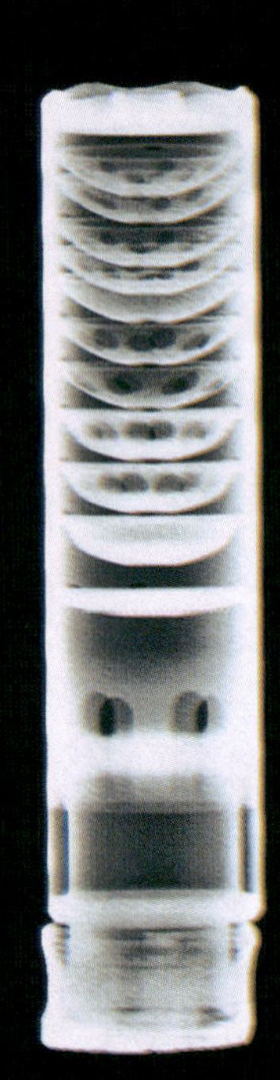

Surefire SOCOM556-RC2

Make: SureFire
Model: SOCOM556-RC2
Weight: 17 ounces
Length: 6.4 inches
Diameter: 1.5 inches
Caliber: 5.56
URL: surefire.com

When someone thinks about SureFire silencers, this is probably the one on their mind. There isn't another can more classic GWOT than the RC2.

Dead Air Nomad

Make: Dead Air
Model: Nomad-30
Weight: 14 ounces
Length: 6.5 inches
Diameter: 1.735 inches
Caliber: .30
URL: deadairsilencers.com

Each baffle of the Nomad sports its own secondary chamber to store and then bleed gas after the projectile passes, effectively making this a silencer with multiple bore evacuators.

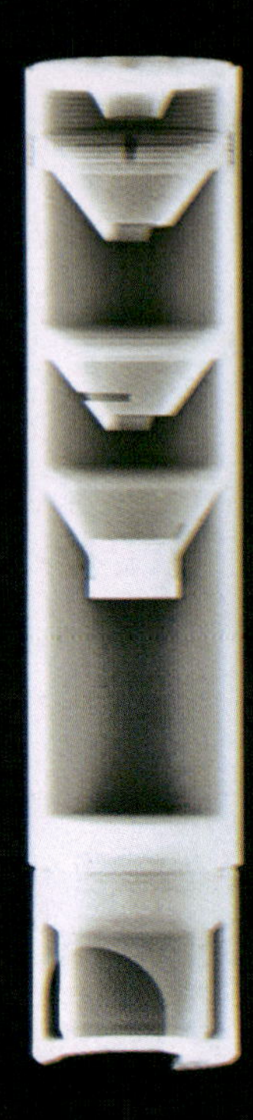

Gemtech GMT-Halo

Make: Gemtech
Model: GMT-Halo
Weight: 12.2 ounces
Length: 6.9 inches
Diameter: 1.5 inches
Caliber: 5.56
URL: gemtech.com

The Ti update to the classic Halo, the GMT-Halo is under half the weight of its heavyweight predecessor.

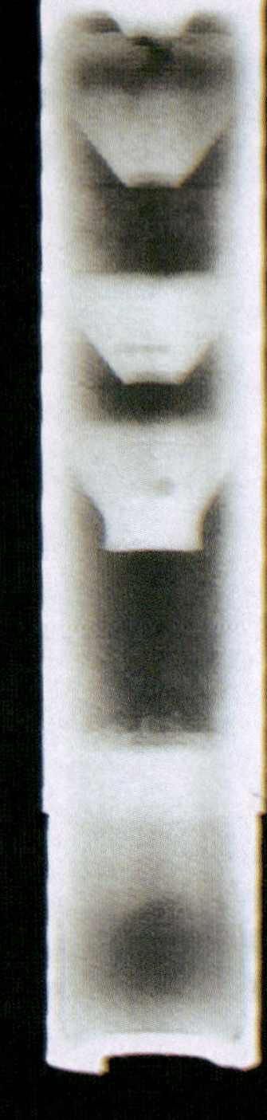

Gemtech Halo

Make: Gemtech
Model: Halo
Weight: 25 ounces
Length: 7.75 inches
Diameter: 1.5 inches
Caliber: 5.56
URL: gemtech.com

Originally designed for use with machine guns (which is why it's so hefty), the standout feature of the Halo has always been its ability to mount to any 22mm NATO muzzle device.

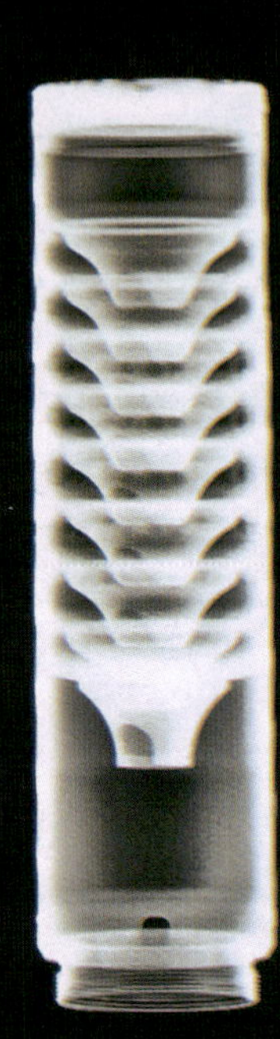

SilencerCo Chimera

Make: SilencerCo
Model: Chimera 300
Weight: 20.1 ounces
Length: 6.9 inches
Diameter: 1.6 inches
Caliber: .30
URL: silencerco.com

The Chimera is one of SilencerCo's best-kept secrets. SilencerCo challenged us to break it, and it's still ticking. This is an excellent choice for AKs in particular.

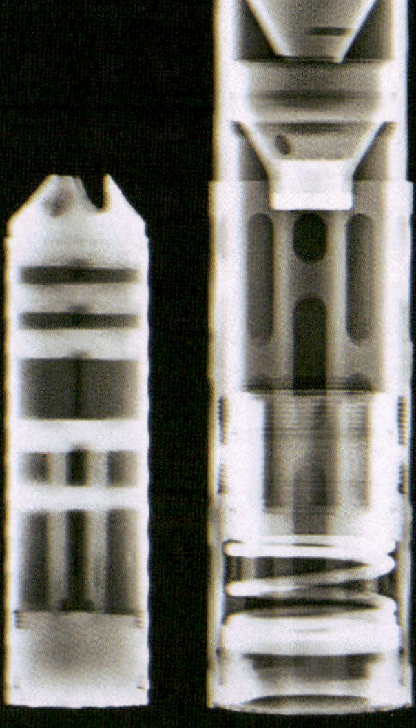

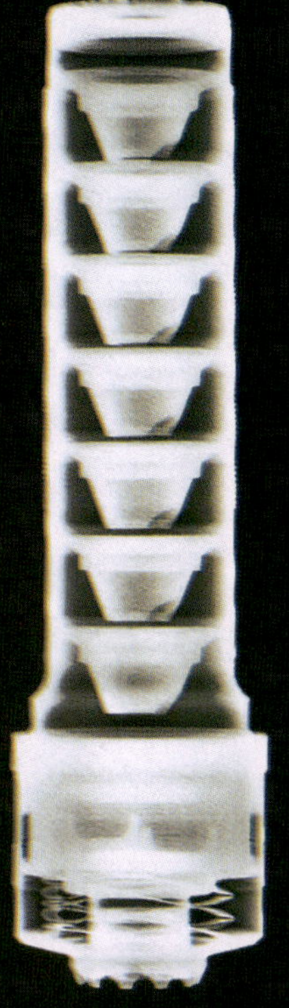

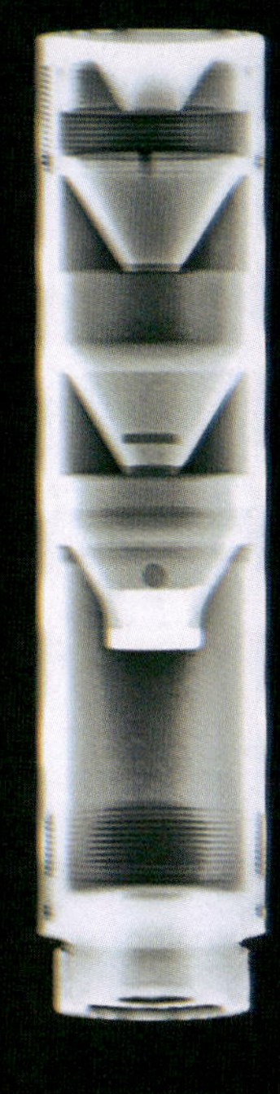

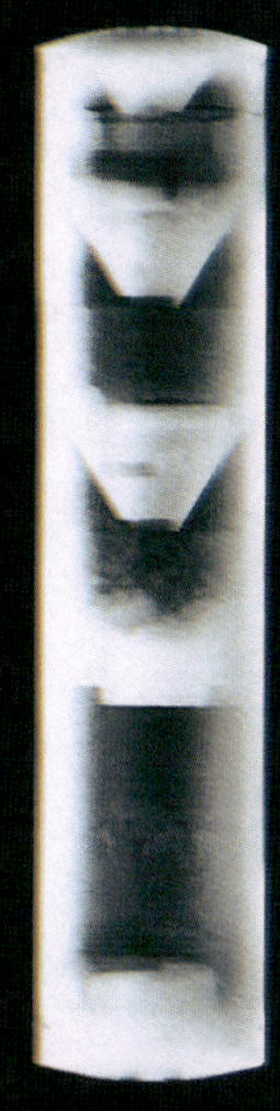

Witt Machine Canooter

Make: Witt Machine
Model: Canooter Valve
Weight: 8.4 ounces
Length: 3.9 inches
Diameter: 1.125 inches
Caliber: 5.56
URL: wittmachine.net

A kissing cousin to the original XM177 moderator, it takes some edge off and changes tone, but don't expect miracles with 5.56. We really liked it with 22WMR.

Gemtech ONE

Make: Gemtech
Model: ONE
Weight: 16.3 ounces
Length: 7.5 inches
Diameter: 1.625 inches
Caliber: .30
URL: gemtech.com

The Gemtech answer to a universal silencer; it was one of the first commercially available rifle silencers with a modular mounting base.

Dead Air Wolverine

Make: Dead Air
Model: Wolverine
Weight: 19.8 ounces
Length: 7.4 inches
Diameter: 1.5 inches (tube), 1.93 inches (base)
Caliber: .30
URL: deadairsilencers.com

Specifically designed for AKs, the Wolverine uses a modified Sandman-S design but with classic styling and baffle apertures that get increasingly wider to make up for nonconcentric threading.

CRUX Ark Neo

Make: CRUX
Model: Ark Neo
Weight: 12 ounces
Length: 7.5 inches
Diameter: 1.5 inches
Caliber: .30
URL: n/a

Though CRUX is no more, what impressed us about the the Ark Neo was the repeatability of their factory Precision Mounting System (PMS).

GSL SWAT-5

Make: GSL Technology Inc
Model: SWAT-5
Weight: 13 ounces
Length: 6.3 inches
Diameter: 1.5 inches
Caliber: 5.56
URL: gslitechnology.com

Unless properly mounted and torqued, direct-thread silencers will sometimes get loose. GSL developed and patented a silencer retention system that makes mounting it both easy and idiot-proof.

Griffin M4SDK

Make: Griffin Armament
Model: M4SDK
Weight: 14.5 ounces
Length: 5.75 inches
Diameter: 1.5 inches
Caliber: 5.56
URL: griffinarmament.com

The Griffin Armament M4SDK uses the SDQD system, which accommodates A2 muzzle devices as well as Griffin's own M4SD II series mounts. It's fast on and fast off.

Gemtech M4-02

Make: Gemtech
Model: M4-02 Piranha
Weight: 16 ounces
Length: 6.25 inches
Diameter: 1.375 inches
Caliber: 5.56
URL: gemtech.com

If you're paying attention, you'll notice a lot of companies using a variation of this baffle arrangement. The M4-02 Piranha proved that just adding more baffles wasn't always the answer.

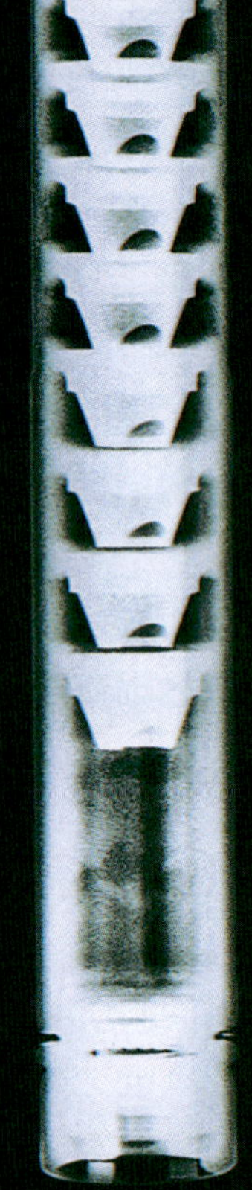

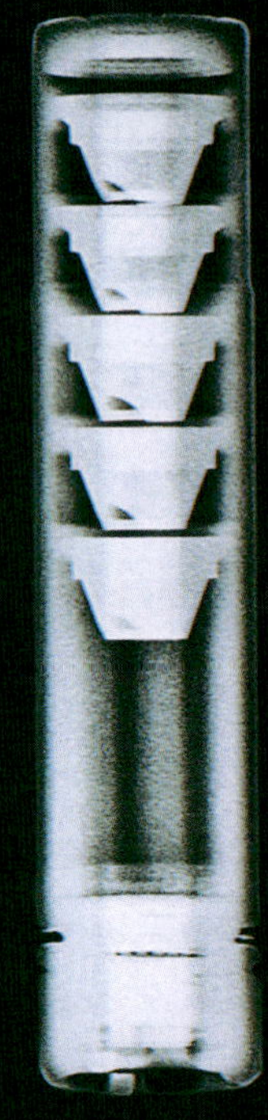

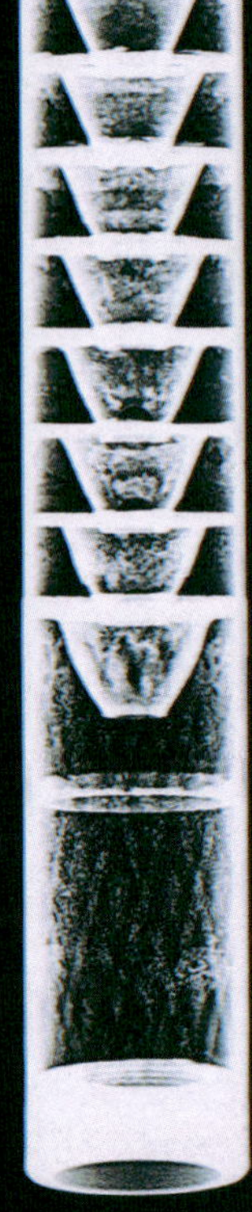

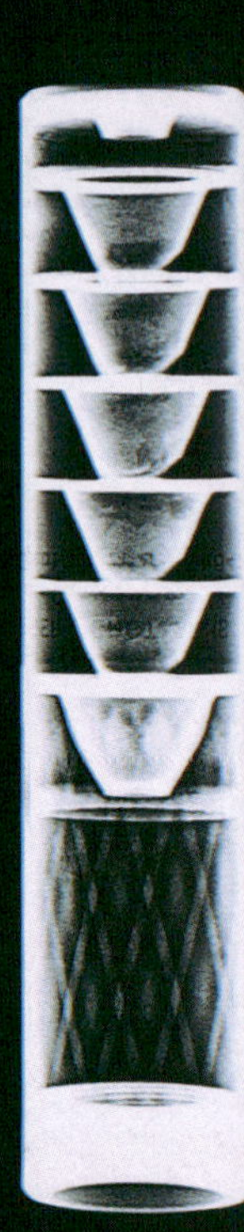

AAC Ti-Rant 9

Make: Advanced Armament Corporation

Model: Ti-Rant 9

Weight: 8.6 ounces

Length: 7.9 inches

Diameter: 1.38 inches

Caliber: 9mm

URL: advanced-armament.com

While no longer a new (or available) product, the Ti-Rant 9 was the lightest silencer of its class in its heyday. Currently, a modular version (Ti-Rant 9M) is available.

Dead Air Wolf-9SD

Make: Dead Air

Model: Wolf-9SD

Weight: 14.7 ounces (full configuration)

Length: 7.58/4.1 inches

Diameter: 1.618 inches

Caliber: 9mm

URL: deadairsilencers.com

The Wolf is fatter than your standard pistol can — because it's made for submachine guns. That doesn't mean you shouldn't try it on your pistol, but you'd better be running a red dot!

Dead Air Sandman-L

Make: Dead Air

Model: Sandman-L

Weight: 21.8 ounces

Length: 8.9 inches

Diameter: 1.5 inches

Caliber: .30

URL: deadairsilencers.com

Dead Air's Sandman series of silencers have been their bread and butter right from the beginning. The L is probably the first introduction most had to their ratchet-on/ratchet-off Keymount locking system.

Dead Air Sandman-S

Make: Dead Air

Model: Sandman-S

Weight: 17.7 ounces

Length: 6.8 inches

Diameter: 1.5 inches

Caliber: .30

URL: deadairsilencers.com

As you can see, the Sandman-S is simply a shorter version of the L. However, on gas-operated firearms, it's nearly as quiet at the shooter's ear due to reduced port noise.

Q LLC Half Nelson

Make: Q LLC

Model: Half Nelson

Weight: 12.2 ounces

Length: 6.85 inches

Diameter: 1.75 inches

Caliber: .30

URL: liveqordie.com

Lightweight, fully welded, tubeless, and all-titanium, the Q Half Nelson uses large, steep, and deeply nested baffles.

TBAC Ultra 9

Make: Thunder Beast Arms Corporation

Model: Ultra 9

Weight: 11.9 ounces

Length: 9 inches

Diameter: 1.5 inches

Caliber: .300 RUM

URL: thunderbeastarms.com

Go to any precision rifle match, and you'll see more Thunder Beast silencers than any other by a fair margin.

TBAC Ultra 7

Make: Thunder Beast Arms Corporation

Model: Ultra 7

Weight: 9.7 ounces

Length: 7 inches

Diameter: 1.5 inches

Caliber: .300 RUM

URL: thunderbeastarms.com

While louder than its bigger brother, the Thunder Beast Ultra 7 is the can for those who want to keep the overall length down on their already-long precision rig.

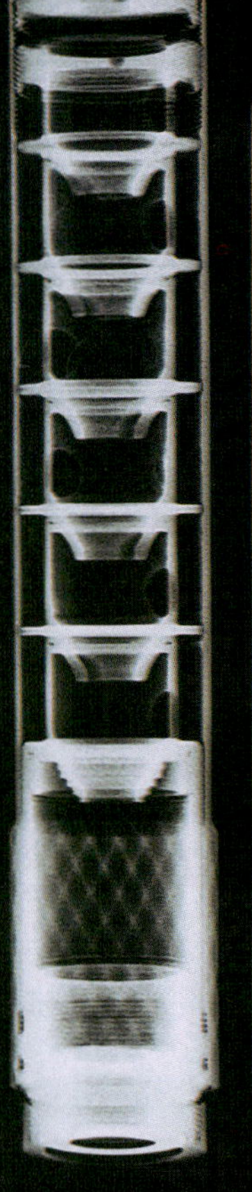

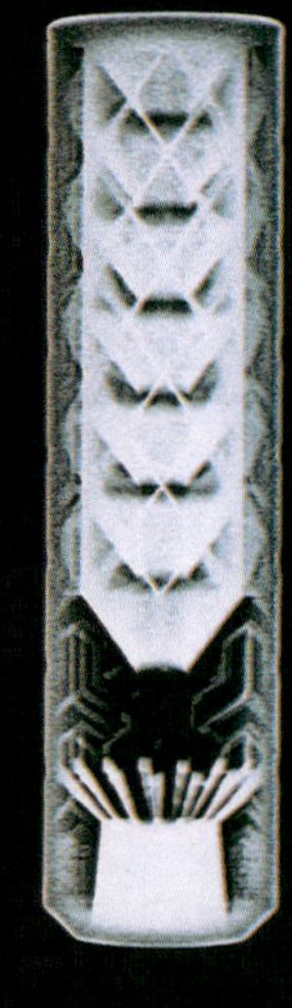

Silent Legion SL-MAG

Make:
Silent Legion

Model:
SL-MAG

Weight:
15.1 ounces

Length:
8.8 inches

Diameter:
1.5 inches

Caliber:
300 Win Mag

URL:
silentlegion.com

It's a 300 Winchester Magnum silencer that weighs under a pound. Though one of the longer silencers on this list, 300 Win Mag requires a helluva lot of space for gas expansion.

Surefire Genesis 762

Make:
SureFire

Model:
Genesis 762

Weight:
17.8 ounces

Length:
8.1 inches

Diameter:
1.5 inches

Caliber:
.30

URL:
surefire.com

Unlike most SureFire rifle cans, the Genesis 762 is a direct-thread silencer. With a body made of stainless steel and an Inconel baffle stack, it's likely this silencer will outlast a barrel.

Surefire SOCOM762-RC2

Make:
SureFire

Model:
SOCOM762-RC2

Weight:
19.5 ounces

Length:
8.4 inches

Diameter:
1.5 inches

Caliber:
.30

URL:
surefire.com

SureFire designed all of their SOCOM-series 7.62 silencers to attach both on 7.62 and 5.56 muzzle devices, but not the reverse. Do you want to push a .30-caliber projectile through your 5.56 can? Likely not.

TDS Bantam II

Make:
Thermal Defense Solutions

Model:
Bantam II

Weight:
8 ounces

Length:
3.8 inches

Diameter:
1.2 inches

Caliber:
5.56

URL:
thermaldefenseinc.com

TDS suppressors are not only 3D printed, their designs cannot be reproduced with standard machines. It's a very impressive silencer for the size.

TDS STRIX II

Make:
Thermal Defense Solutions

Model:
STRIX II

Weight:
16 ounces

Length:
8 inches

Diameter:
1.68 inches

Caliber:
300 Win Mag

URL:
thermaldefenseinc.com

The bigger brother to the Bantam, the STRIX is fatter in all dimensions but uses the same manufacturing and baffle technology.

Surefire FA556-212

Make:
SureFire

Model:
FA556-212

Weight:
16 ounces

Length:
6 inches

Diameter:
1.5 inches

Caliber:
5.56

URL:
surefire.com

Now a legacy silencer, the SureFire FA556-212 was originally named the FA556-K because it's a shorter version of the SureFire FA556AR.

TITANIUM DREAMS

The JK Armament SBR PRO "Solvent Trap" and the Bad-Idea Oil Filter

By Dave Merrill

Just like nearly everything else in this issue, we went the DIY route with our silencers. Normally when you purchase a suppressor, it'll sit at your dealer for months — sometimes more than a year — before the federal government approves your paperwork. But it's faster if you make your own; in fact, we did just that in RECOIL Issue 46.

"Solvent trap" is the current underground moniker for parts and kits used to manufacture your own silencer with an approved Form 1. These are perfectly legal in most jurisdictions, provided you haven't drilled holes in anything just yet. Why would you want to fill out a Form 1 instead of a factory Form 4 transfer? Time, mostly. As we write these words, an online Form 1 takes an average of three weeks whereas even the fastest Form 4 transfer now sits between three and four months. Cost savings is another measure, as most solvent trap kits are available at a significant discount relative to factory silencers (invariably for a reason).

FILTERING "OIL"

We'll get the only-worth-it-for-memes alternative out of the way first: the oil filter adapter.

About a decade ago, we first laid eyes on oil filter adapters; they were originally made by Tom Cole of Cadiz Gun Works and called "solvent trap adapters." Cole tells us, believe it or not, that the original intention was to actually trap solvents. Later, they released their Econo-Can, a registered adapter that's transferred like a standard silencer.

But this type of filter being featured in the 2019 release of *Call of Duty: Modern Warfare* really brought them into the world of mainstream memes. While you can find sketchy adapters on Amazon and Wish, Cadiz Gun Works actually manufactured and legally registered oil filter adapters as silencers and even have an oil filter replacement service for when they wear out.

We teamed up with an SOT to legally register our own. Engraving actually took more work than we anticipated because our adapter didn't have enough surface area. A stainless steel washer was welded to the base for this purpose.

JK Armament has everything you need sans a vise and cordless drill to make an SBR-ready modular titanium silencer.

So, how effective is it? Setting aside that it makes most pistols a single-shot affair because there's no way to attach a booster, the ungainly size that blocks sights on both rifles and pistols and the massively increased weight, simply put: It's not all that quiet. An oil filter silencer essentially is a simple expansion chamber that works for a few shots before having to be replaced. And the larger the bore, the faster it loses any efficacy.

And oh yeah — you can't even legally swap out that oil filter unless you're a manufacturer.

In a world where silencers are OTC just like a receiver, rifle, or pistol, these oil filter silencers would make for a fun party trick, similar to taping a soda bottle to the end of your .22LR. But since legally manufacturing or purchasing adds on $200, and you can't even swap the filter yourself? Hard pass.

JK ARMAMENT SOLVENT TRAP

This brings us to the exact other end of the DIY spectrum, the JK Armament.

JK Armament is a new company, just releasing "solvent trap" kits for sale in 2020. Though the company is new, the people behind it aren't. The namesake of the company, Jake Kunsky, has over two decades of experience in industrial manufacturing and weapons design. Kunsky's company is the culmination of his experience developing products for NEMO Arms, Gemtech, and Maxim Defense. His background with silencer design and engineering gives him unique insight into what today's customers are looking for in a user-configurable and maintainable solvent trap.

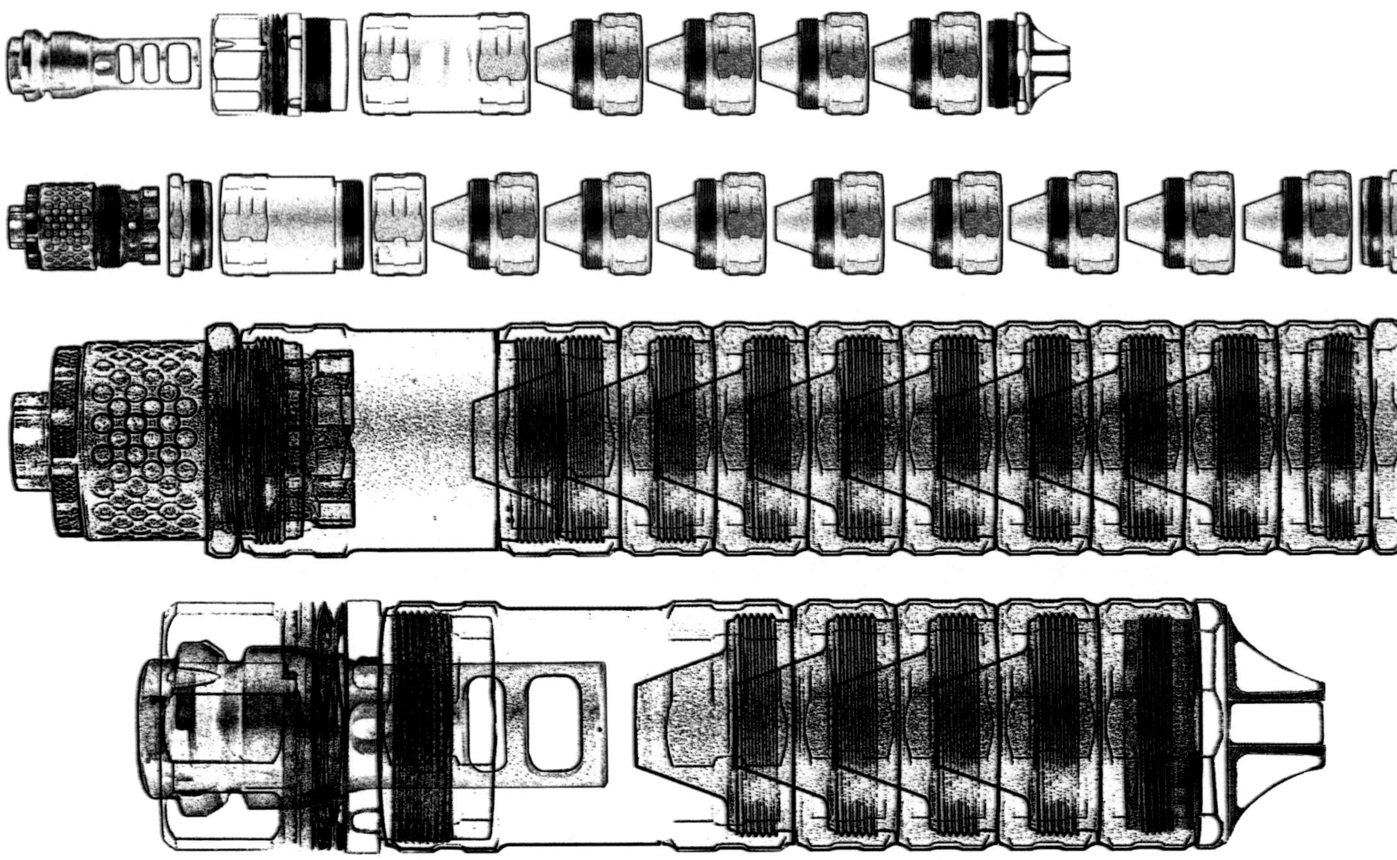

DESIGN

JKA's first product was the JK 155 Modular Solvent Trap (MST) Kit made for either rifles or pistols. The rifle kit is nearly identical to the pistol, except the tube is one piece instead of two, it comes with a fixed barrel adapter, and the eight stackable baffles are optimized for the higher pressures of rifle rounds, so they can't be drilled so wide.

The name of the game is modularity, and it's certainly evident here. Each baffle section threads right on top of another, closely mirroring what we've seen from the likes of Schultz & Larsen for the last decade, the Erector from Q LLC, the Dead Air Odessa, Rebel Silencers SOS, half the product line from Torrent Suppressors, and the new Sig Sauer Mod X.

No need for a tube, spacers, or math, nor do you have to worry about dubious construction, as you would with a cheap Chinesium kit.

You can figure whatever length, caliber, or muzzle attachment you want. Dead Air KeyMo? Sure thing. Pistol boosters? Also, yes.

However, these first kits were made from aluminum. Aluminum is just fine for rimfire, most pistol calibers, and subsonic 300BLK, but once you start rolling out higher pressures, the aluminum will quickly erode away.

Now, JKA does what we anticipated from the very beginning, making them out of titanium. Aside from being turned from titanium, the JK 115 SBR Pro kit comes with four baffles instead of eight, and a new flash hider endcap is available. If you want a larger kit with more baffles, you can simply add them prior to manufacturing.

The base tube and all baffles are threaded HUB 1.375x24 TPI, so mounts will never be an issue as nearly two dozen companies on the market currently make parts and pieces for the HUB. For pistols, a SilencerCo, Torrent, or Griffin Armament 1.375 booster will nest inside, and JKA makes an adapter for those with Gemtech, SDTA, or Liberty boosters.

PLANNING

Planning ahead will help you maximize the utility and versatility of your silencer by making it multi-caliber AF. Not only can you buy the kits, but you can buy individual and incomplete baffles at the same time. Want to make a silencer with 10 baffles? You can do that. Want to make your own bastard version of a SilencerCo Wizard Staff? Sure. Want to drill baffles with different aperture sizes in order to make this a modular silencer that's always caliber-specific? Oh yes, you can do that too.

There's some word going around that you can't register a silencer in a smaller diameter than the bore holes, but that's hogwash. There's no regulation or dictate about how large or small each aperture has to be for a given caliber, so long as

SOCOM Suppressor Firing Table		
Magazine #	Rate of Fire	Rounds Fired
1	(1) round per second	30
2	(2) rounds per second	30
3	(1) round per second	30
4	(3-5) round bursts	30
5	(1) round per second	30
6	(2) rounds per second	30
7	(1) round per second	30
8	Full-auto mag dump	30
	Total Rounds:	240

you can put the registered caliber through it in any configuration — just no extras and no spares. Of course, all of this may be subject to change at the whim of the BATFE.

The Enfield Rifle Co Novus, as seen here in RECOIL Issue 47 also works on this principle with their nesting baffle stack; it's a 5.56 silencer that can be assembled to effectively suppress .30-caliber or .45ACP.

There's currently no reason why you couldn't make a JKA kit with 12 baffles with three different caliber apertures, but we'd recommend Cerakoting or otherwise marking each caliber so you don't inadvertently get a strike.

MANUFACTURING

After we planned out our silencer and received our approved stamps, it was time to hit the workshop. The required info was lasered onto the main tube, so we needed to drill the baffles. Kunsky actually advises against using a drill press, even if you have access to one.

JKA solved any centricity issues with a custom drilling jig. It's meant to be used with a hand drill and a vice — no need for a mill here; the undrilled baffles simply thread onto the jig. A caliber-specific bit available from JKA aligns with a bushing on the jig to ensure it hits center.

The same jig also works for endcaps, and since the jig itself is made of aluminum, your vice can squeeze it hard enough that the endcap won't rotate out when you drill.

With the jig at the ready, along with a cordless drill at a low-speed setting and some machine oil, it was fairly easy. If you drill too fast or don't use cutting oil, you can wreck one of these by chipping it if you're not careful. Take your time, and you'll be fine.

After each new section, we threaded the completed baffles on top of each other to confirm alignment. Make sure you clean the baffle well, or else an errant chip may mislead you.

AT THE RANGE

The original JKA kit was aluminum, relegating it for us solely to pistol calibers and limited use with 300BLK. But with the titanium PRO kit, we knew we could press it much further. And press it further we did.

Using a select-fire 10.3-inch rifle, we decided not to run one or two, but three SOCOM Table II tests. Table II specifies eight fully loaded

After 3 SOCOM tables, the blast baffle (left) looks a bit sandblasted, but you can just rotate it away to increase longevity.

Bits and jigs are available for 5.56, .30-caliber, and 9mm.

blast baffle to the end of the stack to easily increase the longevity of your silencer. And in the worst-case scenario? Have an SOT replace one of the baffles ($125 for a Ti cup, $45 for aluminum).

The flash hider endcap performed exactly as expected too, which didn't surprise us because it's definitely on rather long. For our next build, we'll probably trim them down a bit to reduce the overall length.

30-round magazines with rates of fire ranging from one round a second to full-auto mag dumps. The entire table is performed with no rest between stages, but everything should come down to ambient temperature prior to repeating the table.

Effectively, this is a stress test for the silencer. This accelerated wear gives you a pretty good idea of how the silencer will perform long-term.

Heat is the bane of titanium. Heat erodes titanium, and you can see it happen in real-time when sparks shoot out of the end of your silencer. There was a time when no one would have seriously considered titanium for anything other than a precision rifle or rimfire silencer, but that was nearly 15 years ago.

All said and done, we fired 720 rounds with this test, and while there was noticeable wear on the blast baffle, the others were mostly fine. But, since this is a modular can, you can simply rotate your

LOOSE ROUNDS

You'd be well-advised to add some anti-seize to the threads to aid in disassembly and reconfiguring your can. If you carefully plan your build in advance, you'll have a handy silencer that can effectively eat all sorts of ammunition without a compromise like an oversized bore.

When we're dealing with a new company and a new product, there are typically some warts here and there. Kunsky's experience definitely shows in this design, bringing a lot of advanced features to a home builder that would otherwise require expensive equipment or regaling yourself with eBay parts.

We can't wait to see what comes from JKA next.

OPERATING WITH QUIET IMPUNITY

For over 38 years, 9 patients in suppressor technology, the Founder, Greg Latka has been manufacturing

REX SILENTIUM MOD X

THE SHORTEST AND SMALLEST CONFIGURATION AVAILABLE ON THE BASE MODULE IS THREE BAFFLES. THIS NANO-SIZED UNIT MEASURES 4.1 INCHES LONG AND WEIGHS 8.2 OUNCES.

Sturdy Cans and Boutique Service

By Tom Marshall

Rex Silentium is a small company making a big splash in boutique suppressor service. We recently had the chance to test their MOD X rifle suppressor, and while we're very happy with its performance, we're especially intrigued by their approach to customer service and customization.

Rex takes a very meat-and-potatoes approach to can construction. They start with 17-4PH stainless steel, which is then hardened to 45-47Rc. The baffles are CNC-machined and welded together for a solid, tubeless construction rated for a tensile strength of 190,000 psi. From this foundation, the MOD X is grown into a tough, versatile can with some well-thought-out options for customization so you get exactly the can you need.

The MOD X is a two-part design, with a base unit and a module. You can order it directly from their website in a couple of prepackaged bore sizes and a "standard" configuration of 12 baffles split in half — a six-baffle base unit with a six-baffle extension module, which simply threads on to the base. Off-the-shelf calibers are .308, .338, .358, and .458, which Rex says are their most popular requests. But this is where "standard" ends for the MOD X.

To truly get the most out of this can, the key is to find a Rex Silentium dealer. The MOD X can be built with as many as 14 baffles, split between the two segments in (almost) any way you want. They only do direct sales in the four configurations listed above. But if you place an order with one of

Above: Rex Silentium engraves their suppressor at the base of the tube to avoid having to replace the entire can in case of a baffle strike or re-core.

Left: The beauty of the MOD X is its ability to be customized into two segments with dozen of "split" combinations to create a can that's precisely tuned to fit your desired application and size profile.

their dealers, you can specify total number of baffles, how they're split, and the bore size. If you have a specific length or weight target in mind for your rifle, you can tinker with your configuration to meet these. The shortest and smallest configuration available on the base module is three baffles. This nano-sized unit measures 4.1 inches long and weighs 8.2 ounces. From this building block, each additional baffle adds a half-inch of length and 1.3 ounces to the overall dimensions of the can — all these details are on the Rex Silentium website, giving you a *de facto* recipe book to calculate the exact size and weight of your MOD X in both short and long versions.

Then, there are the bore sizes. The MOD X product page lists 89 separate calibers available for order, ranging from .17 Hornet up to 7.62x54R, with some oddballs in between like 7.7x58 Arisaka, .303 British, .30 Carbine, and .300 H&H Magnum. Don't see your desired chambering? Ask your dealer, and they'll check with the factory for a custom caliber/bore size option. Oh, and Rex's website advertises "no barrel length restrictions."

Our test sample is a .224 bore, tailored specifically to live on a 12.5-inch AR set up as a general-purpose "mini-Recce" rifle. In order to experiment with Rex's minimum-length setup, we ordered a seven-baffle can, in a ¾ split — a three-baffle base with a four-baffle extension module. How much noise reduction do you get from three baffles? Our first test fire on this can took place in a two-lane "pop-up" range at our local FFL. The lane dividers are stamped, diamond-plate metal, like the toolbox in the back of a pickup truck. It's definitely not the quietest can you'll fire, and it's definitely not hearing-safe on paper. But in this environment, the three-baffle base unit of our MOD

X still allowed us to have a conversation with the range officer while shooting. We were concerned a can this small on a 5.56mm SBR would be little more than a blast diffuser. But that's definitely not the case, and the sound reduction is noticeable. Attaching the additional four baffles provides an even more pleasant shooting experience. We like this particular setup because the tiny base unit can live on our shorty full-time to save your hearing in an emergency-use situation while adding minimal length and weight. The full-length extension can be added on during classes and range days for maximum quiet. Want less noise? Add more baffles.

There are a few other considerate features to the MOD X, beyond the extensive degree of customization. Rex Silentium suppressors are engraved at very back of the suppressor body. This way, if you ever suffer a baffle strike, just send the can back to Rex and they can repair/re-core it "in a matter of days." The two halves of the MOD X thread together with built-in witness marks that serve as visual "GO/NO-GO" indicators when coupling the halves. Base modules feature a 1.375x24 HUB mount, making them compatible with a variety of aftermarket quick-detach systems.

The MOD X retails for just under $1,000 — not the most inexpensive can we've come across, for sure. But the hefty construction and intensive customization available at the consumer level make it a not-unreasonable price point for that level of customer service. To spec out your own MOD X, contact your nearest Rex Silentium dealer. R

Right: The MOD X can be quickly set to "short" mode by unscrewing the two halves and replacing the end cap to the base module.

Below: All Rex Silentium suppressors use the industry-standard 1.375x24 HUB mount making it compatible with most existing QD systems.

TACTICAL
LRTS
ELITE TACTICAL
GGP
7.62 X 51

TOP OF THE CLASS

CGS Group Breaks New Ground with the Hyperion QD 762

By Dave Merrill

"We're not going to get lightbulbs from candles if everyone is making the same stuff," said Bobby West, founder and owner of CGS Group. "Everyone doing the same designs the same way doesn't push the ball forward." Right in the middle of our first conversation about new engineering tools, manufacturing, materials science, and, importantly, silencers, West managed to distill his philosophy of life, and by extension his company, into two sentences.

West founded CGS Group at the tail end of 2014, and their first product was the opposite of quiet — the Pandora detonator, most commonly used for breaching charges. It would be 2017 before their first suppressor, the earliest example of the Hyperion, would go through testing for a USASOC small-frame sniper rifle project. Though they didn't end up with that contract, it solidified their decision to continue their work with silencers. That original Hyperion, though not without its warts, would be the basis of all that came after for CGS.

HOW IT'S MADE

CGS Group isn't shy about using exotic materials. Their first foray into this arena was with the carbon-fiber tubed 22LR Siren silencer. With the Hyperion QD 762, we're not talking exotic materials so much as a more-exotic manufacturing method.

The Hyperion QD 762 is manufactured using direct metal laser sintering, or DMLS. This amounts to a Grey Poupon way of saying "3D printed." Through the years, we've seen other silencers made with additive manufacturing, but many of those designs would translate just as well to traditional machining methods.

3D printing can be more than CNC-in-reverse — designs that were either difficult, expensive, or impossible to machine can all become a reality. West explained, "With subtractive manufacturing, you're limited by the confines of your material. With additive manufacturing you're only limited by your imagination."

For example, in RECOIL Issue 43, we covered the Thermal Defense Solutions Bantam-II, an Inconel 3D-printed suppressor. While it's technically possible to machine that design, it would cost in the neighborhood of five figures in machine time per unit, which would only be economically viable for a Russian oligarch.

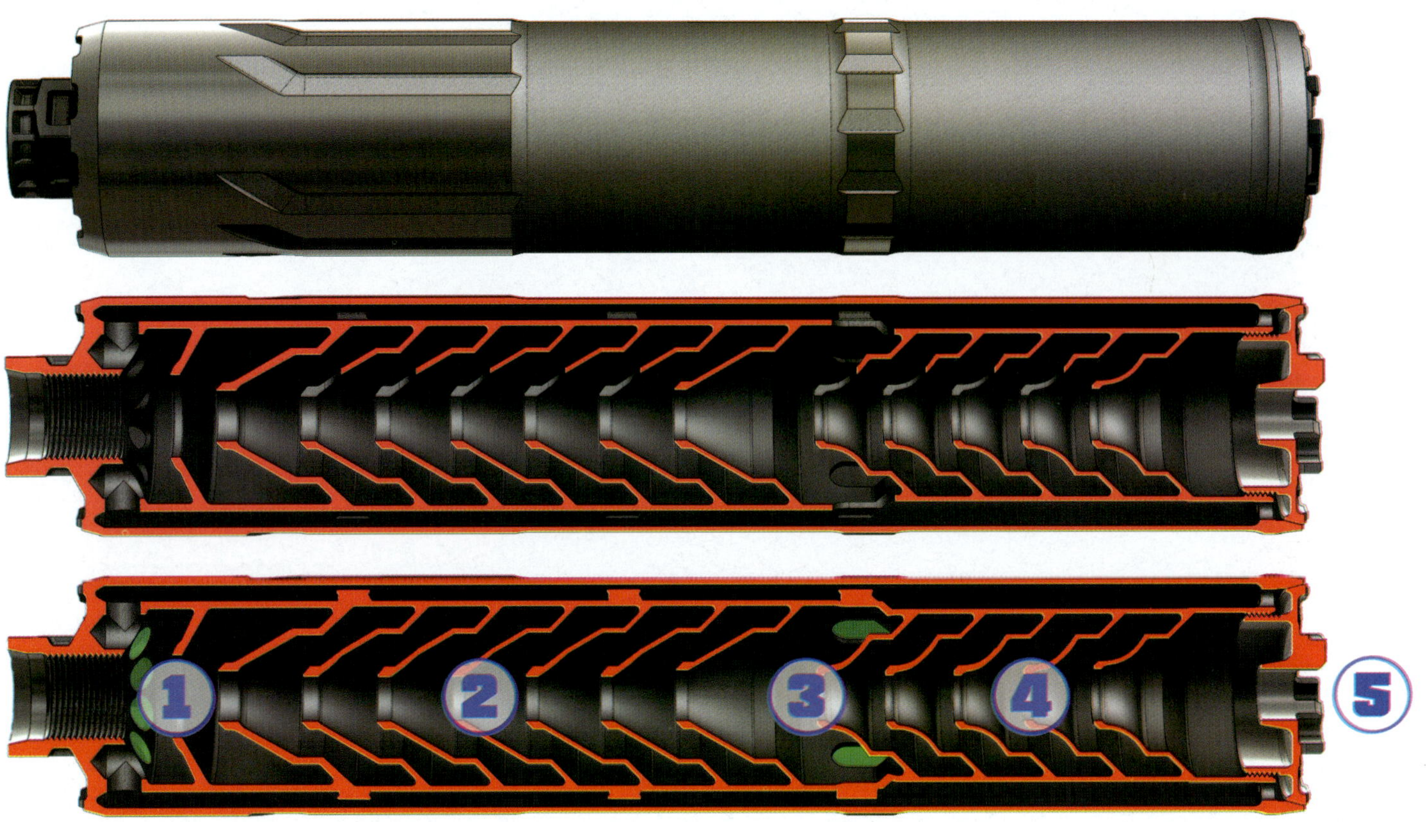

1) Immediately after exiting the muzzle, the gas first hits a sharp baffle to force gas through ports into the exterior annulus/bore evacuator.

2) This section consists of relatively steep baffles designed to accommodate higher pressures.

3) A transition zone, there are ports for gas exchange between the core and the annular cavity.

4) Unlike the baffles in section two, these baffles are more shallow and radiused for lower pressures. Also, note the increased annulus volume.

5) Gas exits the replaceable endcap.

CGS Group isn't using only Inconel; the Hyperion QD 762 is formed from 3D-printed titanium. Titanium and centerfire silencers have had a tenuous relationship through the years. It wasn't so long ago that it would've been considered an inexcusable combination due to how quickly Ti erodes under high heat and pressure. Have you ever shot a titanium silencer and sparks shot out from the end? Those are little pieces of titanium eroding away.

In order to mitigate this damage, starting in 2018, CGS began coating the internals of their silencers with S-Line. Using dynamic compound deposition (DCD) to synthesize boron nitride, we're told S-Line increases hardness erosion protection while decreasing friction. From what we've seen from internal testing at CGS Group, it works well for this purpose. That said, we've seen a veritable laundry list of über-coatings before, some better than others, but keep your eye on this.

DESIGN & OPERATION

Check out the cutaway of the Hyperion QD 762 to get a peek at what's going on here.

Immediately, we can tell this isn't a traditional silencer full of 60-degree cones. There's a cavity space between the baffles and the outer shell of the Hyperion, called an annulus. Though, there's more to it. Firstly, it serves as a bore evacuator, a place to temporarily hold gas that'd otherwise contribute to muzzle blast and sound. Initially developed for large guns such as those on tank or artillery pieces, a bore evacuator is a space where high-pressure gas flows due to an induced pressure differential. Prior to firing, the gas in this cavity is initially the same as atmospheric pressure until the projectile passes. When the projectile exits the silencer, both the vacuum of the wake and the lower pressure outside pulls the gas from the annulus out into the open air.

We first saw this used for suppressors with the Gemtech Integra (see RECOIL Issue 35). Then, Dead Air incorporated a bore evacuator in their NOMAD silencers (see RECOIL Issue 51). CGS went a step further by consolidating this feature and more into one unit.

Let's walk through the Hyperion QD 762 for a bit: Traditional silencer designs have a larger blast chamber in the first several inches of the silencer to accommodate the higher pressures of centerfire rifle rounds. But in this design, the first place the gas hits is a very sharp, shallow baffle. Gas loves to travel in the path of least resistance and will try to move around any obstacle that hinders it. This first flat baffle intentionally increases pressure early on in order to divert some gas into holes immediately following the threads, to the annular cavity where it can be further directed.

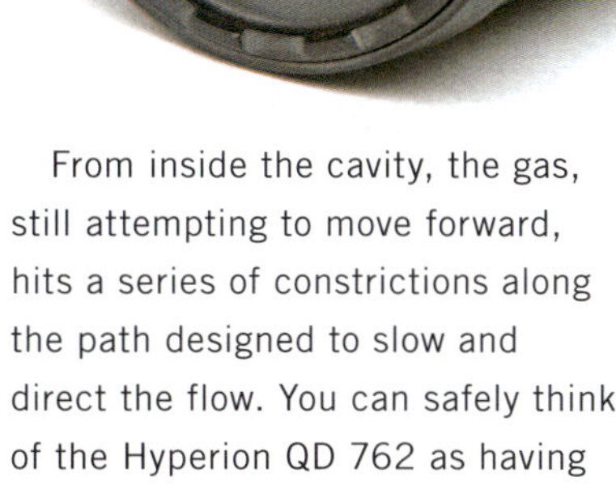

From inside the cavity, the gas, still attempting to move forward, hits a series of constrictions along the path designed to slow and direct the flow. You can safely think of the Hyperion QD 762 as having a core within a core.

This temporarily contained gas won't flow into areas with higher pressures, so it can't freely flow back into the core of the silencer until internal pressure is reduced after the projectile exits.

The next series of seven baffles are intended to deal with the rest of the high-pressure gas put out by your centerfire rifle cartridge. Then, we hit a transition zone about two-thirds through; it's about here where the initial gas pressure starts to drop and the increased volume helps the process along. There are additional ports where gas can be exchanged between the bore evacuator/annulus and the center bore of the silencer.

At this point, the baffles more closely resemble a pistol silencer because these radiused, more-shallow baffles do better with lower pressures. The lower pressures toward the end of a silencer are also the reason why you'll sometimes see silencers use different baffle materials depending on where they're positioned inside the suppressor. Area 419 does something similar with their Maverick precision silencer (see RECOIL Issue 47), except that the Maverick comes in multiple pieces.

It's in this section that the bore evacuator or annular cavity increases in volume, allowing the gas to linger further still before finally exiting the silencer. Finally, everything exits the replaceable endcap. Not only does this mean endcap strikes can be fixed, but gas flow can be further conditioned.

Among all of this new/shiny is a common, but not yet universal feature: wrench flats on *both* ends.

Yes, indeed, there's an awful lot going on here!

QD?

Admittedly, when we first opened the box, we were a little confused. Despite having "QD" right in the name, the Hyperion has what first appeared to be standard 5/8x24mm threads. A quick internet search also showed the silencer listed as "direct thread," so what the heck is going on here?

In short: government contract mucky-muck.

As mentioned, the original Hyperion was developed specifically for a DoD semiautomatic sniper project.

There are generous wrench flats on either end of the Hyperion QD 762.

Most of the silencers submitted for consideration were .30-caliber silencers lightly reworked (or not) for what was at the time 260 Remington (later changed to the nearly ballistically identical 6.5 Creedmoor). West felt he had the best design, with the conspicuous exception of a quick-disconnect muzzle device.

Between direct-thread and QD silencers is an odd in-between space we'll call hybrid mounting systems. These systems usually entail a muzzle device that's affixed to the barrel but has exterior threads (often thick ACME threads) to which the silencer is attached. You're still threading and spinning, just onto the device itself and not the barrel.

The U.S. government considers a suppressor "QD" so long as it can be attached faster than its direct-thread brethren. CGS simply developed a taper hybrid mount with external threads but instead of proprietary threads, they're also standard 5/8x24mm. Since this new mount includes a 25-degree taper, it ends up being a more solid solution than standard direct-thread.

The Grey Ghost Precision GRIM rifle chambered in 6.5CM is similar to the rifle the Hyperion QD 762 was designed for in the first place.

QD by government standards? Yes. Can you use it on your direct-thread Griffin Armament, Q LLC, or SIG silencer that also uses a taper mount? Also, yes.

ON THE RANGE

Since the Hyperion QD 762 was originally designed for a .260/6.5 gas gun, that's exactly what we tested — a Grey Ghost Precision GRIM MKII in chambered 6.5CM served as the host. The added 9 or so inches of additional length makes for a friggin' harpoon of a rifle but less than other precision silencers.

Normally, when we say "gas gun" and "precision silencer" in the same sentence, we're interrupted by a boatload of leaded gas to the face. We didn't have that problem here for two reasons: The Hyperion is designed to be low-pressure for use in semiautomatic rifles, and the GRIM MKII has an adjustable gas block.

On the indoor range, the Hyperion QD 762 sounded similar to an unsuppressed 22LR, with the notable exceptions of a deeper tone and the sound of the much-larger projectile smacking the steel trap. Outdoors? Much of the same — hearing safe from the firing line but more noticeable downrange. It isn't "movie quiet," but it's the best 6.5CM suppression on a gas gun we've ever heard. Excellent, low tone and extremely comfortable without ears — you can use this one all day.

Frankly, this shouldn't be that surprising, as we had a conversation without hearing protection while firing a Barrett .50BMG M82A1 equipped with a CGS Group silencer just a few months ago.

Sure, it increases overall length, but auditory signature was reduced far, far more than a sub-10-inch silencer should be capable of. How about recoil? No perceivable increase.

LOOSE ROUNDS

The majority of silencers on the market are no-new-innovation copies of someone else's more successful product, in the same way it sometimes seems the latest striker-fired wonder-nine is just a variation of a riff on a Glock 19. But that's not what the Hyperion is. While we've seen many of these elements in previous designs, we've never seen a company put them all together in one place before — even if you discount all of their new additions. CGS Group bet they could make something that was more than simply a summation of parts, and they bet right.

They push the limits of what's possible by using all the new tools and tricks modern manufacturing gives them. In the future, we'd love to see a 1.375x24 HUB mounting system to drastically increase mounting options.

CGS Group has set a high bar in terms of both design and manufacturing, and there's no doubt in our minds that several of the features you see in these pages will become standard within the next several years — and our ears will thank us for it.

CGS GROUP HYPERION QD 762

LENGTH: 9.5 inches
DIAMETER: 1.75 inches
MATERIAL(S): Titanium
WEIGHT: 15.1 ounces
CALIBER: Up to 300RUM
MOUNT: Direct-thread
MSRP: $1,379
URL: cgsgroup.com

S
E
F

ELIMIN8TING THE NOISE

Lone Wolf Distributors' New Venture into the World of Silencers

By Alexander Crown

Silencers are awesome, and assuming you live in one of the still-somewhat-free 42 states that allow them, you should own a few. Here, we take a look at the new Elimin8r 45 silencer from Lone Wolf Distributing.

Lone Wolf has been in the Glock accessory game since 1998. They were (and still are) a huge supplier of aftermarket Glock threaded barrels, so their roots run deep with silencers. Frankly, we're surprised it took them this long to make their own, seeing as a silencer is one of the best Glock accessories available. With so many different silencers on the market, the Elimin8r 45 touts features that may not necessarily be unique, but they're definitely desirable for first-time can buyers as well as seasoned NFA enthusiasts.

ANY WAY YOU WANT IT

The Elimin8r 45 silencer is a modular design that allows the user to choose how many baffles they want to use; this is the same method used by the Dead Air Odessa, Q LLC Erector, JK Armament JK155, and similar. The silencer comes outfitted with a main body, eight baffles, and one end cap that adds up to 8.45 inches overall. The baffles are all identical and are clipped, so the user doesn't need to worry about putting them back together in a specific order or setting aside a separate, special blast baffle.

Each baffle also has an O-ring installed, which helps keep the threads clean and makes taking the can apart much easier. The O-ring also assists in preventing the baffles from becoming loose under heavier recoil or high rates of fire. Lastly, the O-rings allow you to align the baffles in a specific way while keeping them snug. For example, if you wanted to ensure all the baffle clips alternate or align to see if that affects either the accuracy or auditory signature, you can do that.

The end cap is sized for use with .45-caliber projectiles, and Lone Wolf plans to produce a smaller 9mm version in the future. We've been told the end cap will accept a wipe, if you're so inclined. For those who aren't familiar with them, many older silencers used wipes made from urethane or similar materials to assist in sound reduction. Wipes wear out with use but can also be very effective. The best current example of wipe usage is found on the Gemtech Aurora II (see RECOIL Issue 37).

By having completely modular baffles, you can decide how much suppression you want; use all the baffles for max suppression or for a more compact package, only use a few. This is a nice option to have when using a silencer on multiple platforms and calibers where subsonic loads, barrel lengths, and other considerations are in play in terms of sound mitigation. It's also important to note that fewer baffles generally result in a larger flash signature as well.

The Elimin8r silencer is constructed from anodized 7075 aluminum, with the piston and spring formed from 17-4 steel. Pistons for use with tilt-barrel pistols aren't included with the can, nor is any thread mount, which is sort of a bummer. Lone Wolf does offer them for sale — and, importantly, pistons from SilencerCo, Rugged, and Griffin Armament are all compatible with the Elimin8r. Lone Wolf's piston offerings are in ½-28 or 0.578-28, so if you want to deviate from those two threads, check the other manufacturers.

There are multiple flats on each baffle to aid in disassembly.

A properly greased booster assembly can keep everything quiet and running.

The main body of the Elimin8r 45 is threaded 1.125-28TPI, making it compatible with other manufacturers' mounts. The Elimin8r 45 is a common 1.375 inches in diameter, so it's compatible with suppressor-height sights on most pistols, and even more so on pistols with a red dot. The silencer is advertised as full-auto rated up to 10mm pressures, without mention of specific barrel lengths. Reaching out to Lone Wolf, we wanted to know if the Elimin8r is good to go on .45-70 and .458 SOCOM; they said on a 16-inch barrel or longer it should be fine. There was no mention of full-auto fire with them, but an automatic .458 SOCOM sounds like a fun, albeit expensive, adventure.

ON THE RANGE

Out on the range we wanted to see, or actually hear, how the can performed on two common calibers: 9mm and .45ACP. What better way to use a Lone Wolf product, known for Glock accessories, than on the back-to-back world war champ, the 1911?

This Remington R1 has been shot almost exclusively suppressed since it came out of the box many years

Each baffle features symmetrical clips and an O-ring.

ago. When testing a new silencer, it's best to use a well-seasoned and reliable firearm. In the event some reliability issues crop up, it makes the problem much easier to diagnose. For .45ACP, we opted to keep the silencer in its full-length configuration. To our semi-trained ears, the silencer performed well at taming the report. Functioning was also no issue. The can did have more perceived gas than others we've shot, but that may be attributed to the lubricant on the piston or the ammunition. On this range trip, we shot PMC 230-grain FMJ.

Next, it felt appropriate to shrink the can down for a smaller gun like the SIG Sauer P365XL, with a Griffin Armament ATM barrel. Five baffles and the end cap seemed like a good size; we would've liked to have a dedicated 9mm end cap, but they weren't available yet at the time of writing. The sound level was adequate using 147-grain American Eagle FMJ, but nothing close to the tried-and-true SilencerCo Omega9k that we regularly use.

The blowback seemed to be less but still present, which could be from the lubricant being blown out previously.

Taking the can apart on the range was very easy, particularly compared to other modular designs we've tested. No tools were necessary.

Finally, we used a Rugged Suppressors 3-Lug mount and mounted the Elimin8r 45 to a select fire MP5 in the full-size configuration. Full auto is always a blast, and the can made it even better. There were no issues with mounting, blowback was practically non-existent, and the sound level was downright pleasant using subsonic 9mm ammunition.

MAKE: LONE WOLF
MODEL: ELIMIN8R 45

CALIBER: .45ACP, UP TO 10MM PRESSURES
LENGTH (FULL CONFIGURATION): 8.45 INCHES
WEIGHT (FULL CONFIGURATION): 10 OUNCES
DIAMETER: 1.375 INCHES
MSRP: $599
URL: LONEWOLFDIST.COM

LOOSE ROUNDS

The Elimin8r .45 is a good general-purpose pistol-caliber silencer. It has decent features, can be user configured for length, and is supported by multiple manufacturers for mounting options. We can't help but chuckle at the name being spelled in Leetspeak. The ability to clean the can makes it ideal for high-round count pistol-caliber carbines, subguns, and anyone who shoots dirtier ammo (like subsonic 300 BLK). We'd like to see Lone Wolf offer more choices for pistons and mounts in the future.

Lone Wolf's first foray into the silencer world is a solid choice, and we look forward to seeing what they bring to market next — hopefully an integrally suppressed 9mm Glock slide, if we dare say so. R

THE HARD PART FIRST

We Check Out Maxim Defense's MSX Silencers, New Rifles, and More

By Dave Merrill

Maxim Defense started as a one-man operation in 2013 with an idea for a new PDW stock. That stock begat a PDW brace, and then acquisition of an awful lot of talent. Three years ago in RECOIL Issue 41, we showed you their PDX, a 5.5-inch barreled PDW — because a PDW stock needs a PDW to live on. But what really impressed us is that the PDX actually functions properly, even chambered in the difficult-in-ARs 7.62x39. At the end of that article, we said there was a lot more to come in the future, and here we are now in the future.

Some of what you'll see in these pages are previews, and most are pre-production prototypes but should be representative of what you'll soon see on store shelves, including new weapons, specialty ammunition to make those weapons more effective, and silencers to go on them. Some names and a weight here or there may change, but the bones are all here. If you're attending SHOT Show in January 2022, you can see all of this firsthand on the show floor yourself.

MSX SUPPRESSORS – MAXIM SILENCERS

Even though part of the company name invokes Hiram Maxim himself, Maxim Defense spent several years to develop their first muzzle mufflers. The official name of the suppressor line from Maxim is MSX Suppressors, but we'll just call them what they are — Maxim Silencers. After

Smith & Wesson acquired Gemtech and moved it across the country, Maxim Defense soaked up much of the talent S&W left behind in Idaho. Once we saw Dr. Phil Dater (see RECOIL Issue 41), grandfather of modern American suppressors, joined Maxim's team, we knew it was just a matter of time before quiet-makers would be on the way. They also worked closely with CGS Group for some Maxim cans.

All told, we've been looking at various different design iterations since before the start of the COVID pandemic.

MXS lineup, left to right: MMG-240, SOC, PRS, DRF-22, and Maxim CGS Micro.

Maxim Machinegun Suppressor: MMG-240

Michael Windfeldt, founder and owner of Maxim Defense, said, "We do the hard stuff first." In the case of suppressors, this refers to successfully silencing a 7.62x51mm FN MAG belt-fed machine gun. We're not talking about attaching a suppressor to a belt-fed as a parlor trick for Instagram points, but as a practical-use military item. The Department of Defense is looking to put suppressors on absolutely everything. The goal is to suppress all of SOCOM weapons by 2026 and the entire military by 2035. That's a heavy lift, but they're taking it seriously. Not only do quieter guns mean sneakier operators and more effective communications, importantly they'll also take a substantial bite out of hearing loss and concussion, with the estimated $75-plus billion paid out every year for hearing claims alone.

We first got handsy with a prototype years ago and couldn't believe the performance: No earpro, indoors, and not uncomfortable. It's hard to overstate how crazy that is. OSHA's "hearing safe" threshold is 140 dB, and the MMG-240 brings a M240 down to just 136 dB. In addition to sound and flash, the concussion was significantly reduced. Continued development saw an increase in longevity and further abatement of noise. We've seen earlier stainless steel units with more than 10,000 rounds through it that are still effective. While you probably won't buy one to pop in the safe at home, it could very well end up in the hands of those wearing uniforms.

The SOC breaks down into four pieces including the mount. The hybrid core has a carbon scraper and an integrated M-baffle.

SOC / PRS

SOC ("Suppressor, Optimized, Carbine") came about from lessons learned in developing the MMG-240. It's what we'd call the main line of Maxim Silencers. We've been privy to the development of

The final details of the SDX uppers are still being hammered out; half of this handguard is printed. You can get a look at the slimmer handguard locking mechanism of the new MDX line here.

the SOC over the last few years, and every iteration has been a little quieter and a little lighter than its predecessor. SOC silencers have variations for caliber (5.56mm, 7.62x39, 6.5CM) and also mounting solutions. There are four main pieces to each SOC suppressor: the mount, serialized unit, hybrid core, and exterior tube. Though Maxim has their own mounts, SOC suppressors are threaded in the HUB-standard 1.375x24 TPI, so there are dozens of mounting options.

Built for durability first, the internals of the SOC line are 17-4 stainless topped with titanium external tubes. At first blush, the left-hand threaded guts look like a normal monocore, but there's a hidden M-baffle. Tucked inside at the end of the blast chamber is a more-traditional baffle that significantly reduces gas velocity as it moves through the core. A sharp-edged carbon cutter cleans and scrapes during disassembly, and the tube secures in place with a ratcheting system.

The SOC units are hefty at 22 ounces ("high durability" is a synonym for "heavy") but can bring an 8-inch full-auto 5.56mm rifle down to 135 dB, while impressively also surviving multiple SOCOM firing tables. "Full-Auto Rated" is nearly always a marketing term, but not when we're talking about actual belt-feds.

The PRS line is virtually the same as the SOC, sans materials. Purpose-built for precision and suitable for hunting, the PRS silencers swap the heavier stainless steel internals for aluminum and titanium. No, you won't want to put this one on a machinegun, but at half the weight of the SOC, you'll be A-OK carrying it up a mountain.

Maxim CGS Micro

As soon as we saw the Direct Metal Laser Sintered (DMLS) exterior of the Maxim CGS Micro, we knew CGS Group was behind it. This squat, direct-thread silencer is just under 5.25 inches long and adds just 4.6 inches to the overall length of the host. Entirely constructed of 718 Inconel, the Maxim CGS Micro is very tough and weighs in at 15.7 ounces — it's very dense.

The Maxim CGS Micro is essentially a CGS Hyperion QD (see RECOIL Issue 55) that's been squished down to fit. You can't expect full-size performance from a squat little can like this, but it effectively takes off the sharp edge of concussion and brings the blast down to a more manageable level.

SDX

The SDX is the integrally suppressed carbine line and will be available as stand-alone uppers and complete weapon systems. It'd be more accurate to say these are integrated rather than integral silencers, but they aren't just a can-on-a-stick. Nested silencers can become untenably hot with long strings of fire, so Maxim decided against docking the can inside the handguard. The uppers are tuned for the given caliber and suppressor, and with that comes an increase in performance. A 5.56mm SDX upper clocks in right at 131 dB, plenty hearing-safe.

Rimfire

The DRF-22 is the current rimfire offering, and at 3.4 ounces it's extremely lightweight. We don't have

The only people who don't like hush puppy 22s are those on the receiving end of them. Though the cores of the DRF-22 and the MKIV-SD are the same, there's more going on under the hood of the Ruger.

A peek of what's to come, including a large frame and a rimfire to balance the line.

a ton of time behind it, but most all 22LR silencers sound pretty good, including this one. The star of the rimfire show at Maxim Defense is the integral Ruger Mk IV, MKIV-SD. Yes, you've seen these before, because Dr. Phil Dater has been producing versions for decades, this being his latest handiwork.

Yes, it's murdery, fun, and gets the mind churning. Add an IR laser and some night vision, and it's a real party. The only folks who won't want one of these are those on the receiving end of it.

AMMUNITION

With extremely short rifle barrels come anemic ammunition performance — at least if you're using standard ammunition. While short-barrel-optimized ammunition certainly exists (see CONCEALMENT Issue 11), Maxim wanted house-brand offerings. Working with established manufacturers Fort Scott Munitions and Red Mountain Arsenal, they developed new solid copper spun TUI (Tumble-Upon-Impact) projectiles in various calibers and grain weights, all far less reliant on velocity for their primary wounding mechanisms. This is definitely not bulk blasting ammo, unless your last name is Bezos.

Unjacketed frangible ammunition that's safe to shoot on steel from nearly contact distance will also be available in 9mm, 5.56mm, 300BLK, and 7.62x39. We're told it'll hold together inside a silencer as well; current-issue frangible is semi-jacketed and can be ripped apart by the intense heat and pressure inside a suppressor.

We've peeked behind the curtain of what's coming next from Maxim ammunition, and you'll be impressed.

The frangible ammo (far left) is unjacketed and designed for close ranges on steel.

The TUI loads are designed to perform with the lower velocities of short barrels.

LEXUS & TOYOTA

When companies start really piling on the SKUs it can practically take a decoder ring to decipher. Maxim currently has two broad lines for its firearms, MDX and MD. The MDX is their premium line. Here you'll find the bougie billets and premium machining, HK 416-height uppers and handguards — the sex and sleek for those who appreciate the craftsmanship. MD ("Maxim Duty") guns share all the same internals as the MDX line, but with beer-money receivers and a heavy focus on duty use.

The latest MDX rifles are sleeker than ever, with an additional 10 ounces of metal removed from the receivers.

The premium MDX (top) and the duty MD (bottom). The internals are all the same, but a little less sexy with the MD.

MD and MDX rifles are available in a half-dozen different barrel lengths (from 5.5 to 16 inches), three calibers, and different stock/brace options. Over the years, Maxim has teased different 9mm PCCs, and while prospecting through their engineering office we were teased once again with a new non-firing prototype. It's ambi, locks back on empty, and, yes, it eats from Glock mags.

MD:11, a large frame semiautomatic rifle, is also in the works. Using an SR-25 pattern receiver, the example we shot was chambered in 6.5CM and featured a Proof Research carbon- fiber barrel. It's a safe bet you'll see more about these weapons when they hit production status.

EDUCATION

In addition to all these new products, Maxim Defense also teamed up with established companies to build up the education side of the house — the gun, the ammo it eats, the suppressor that calms it, and the knowledge to use it. For shooting and tactics, Maxim instructors Seaux Larreau and Dennis Bechtel can often be found at DARC, the Direct Action Resource Center located in North Little Rock (see RECOIL Issue 28). On the

An early prototype Maxim SOC suppressor on a PDX.

more academic end of the stick, there's Dr. Phil Dater's three-day ITAR-restricted silencer class. A must for suppressor nerds, this course is on the history and development of silencers, now taught by Maxim Institute's Travis Bundy.

THE FUTURE

We visited Maxim's new facility in St. Cloud, Minnesota, a location chosen explicitly for the high concentration of precision machine shops and industry knowledge. It's here that much of their new products developments have come to be.

We love to see small American businesses grow. Maxim Defense has consistently pushed toward the next logical step in their evolution, either by partnering with an established company or by acquiring the human talent required. And importantly, they've successfully and appropriately scaled. Failure to scale is a common reason for the demise of promising companies; you can have a great, popular product made by good people and still have everything not work out in the end. Whether it's due to lack of funding, mismanagement, or lack of flexibility, not every decent product actually makes it to market alive.

We're always trying to see what's on the next page and scraping the next layer. When we talk about Maxim Defense, something to keep in mind is that they're not just a gun company; they've become a defense development company. We've gotten a glimpse or two of what's next from Maxim Defense and the trajectory they're following. Not bad for a company with a single product started in a garage — what's more American than that?

PROFESSIONAL GRADE EVERYTHING.

FIREARMS
ACCESSORIES
AMMUNITION
SUPPRESSORS

MAXIM DEFENSE®

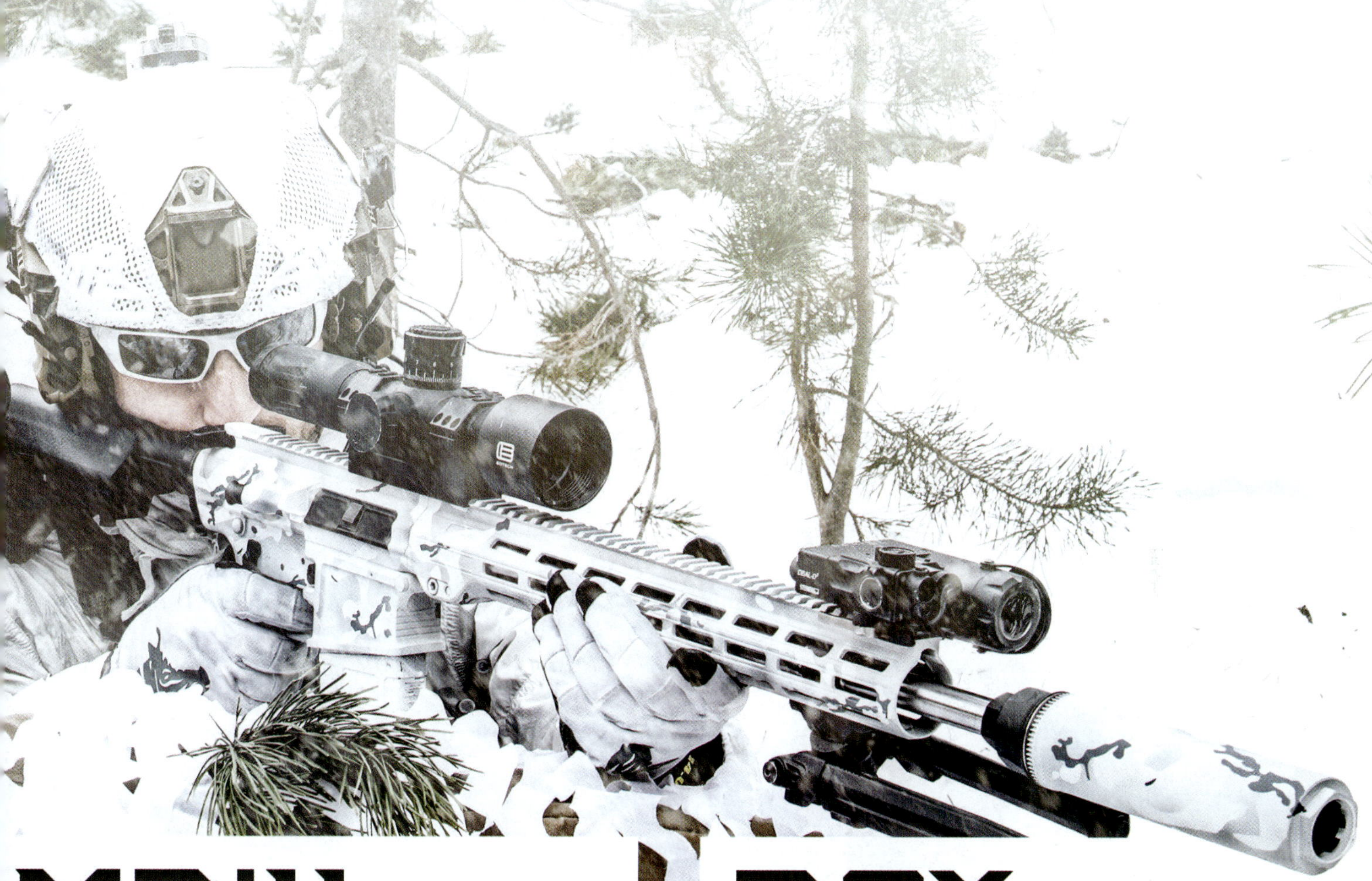

MD:11

CALIBERS: .308 WIN | 6.5 Creedmoor

COLORS: Arid | Urban Grey | Black

Heavy-hitting firepower from Maxim Defense. The SR-25 patterned MD:11 rifle series brings long range accuracy with Maxim Defense durability. Get yours now from select dealers or at MaximDefense.com.

Available with a match grade stainless steel barrel or Proof Research carbon fiber barrel.

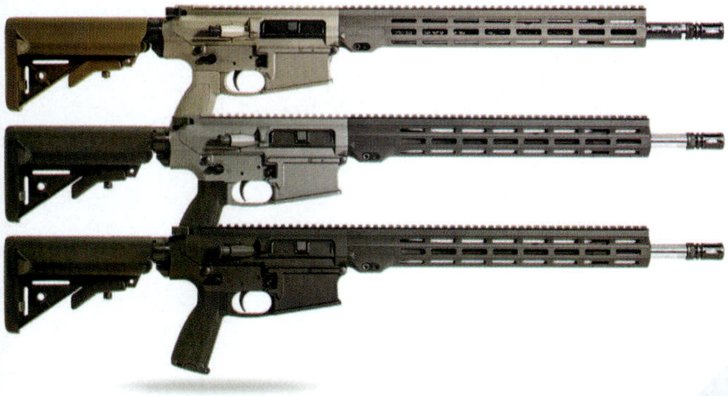

ALL NEW

DSX

SUPPRESSORS

Duty built for the world's most rigorous demands, this suppressor is designed to be the most robust and strongest suppressor in category. It has been repeatedly tested with SOCOM tables across a variety of calibers and barrel lengths, down to 8.5" 5.56 NATO with M855 ball ammunition, and has survived with no damage or significant changes in sound reduction.

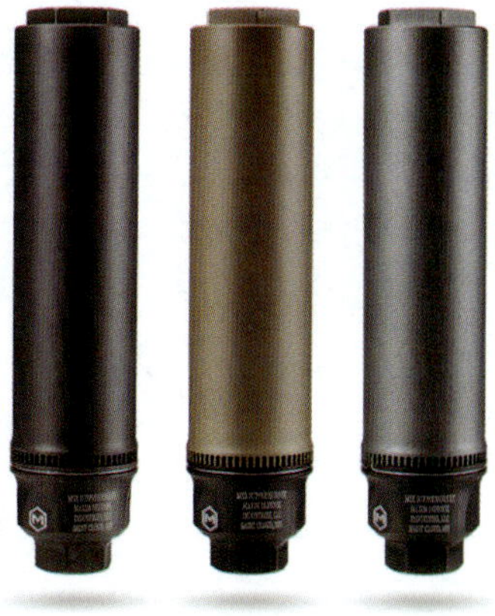

Calibers:
5.56mm
7.62mm
6.5mm

View the full lineup at www.MAXIMDEFENSE.com

HYBRID MOMENTS

Suppress All Things with the SilencerCo 46M

By Mike Searson

There was a time when silencer ownership used to be infuriating. You might want to put a can on every gun you owned, but between different bore diameters, different thread pitches, and the way most silencers were built, it was an exercise in futility as well as a great expense.

Over the decades, this has changed. Advances in materials, designs, and engineering have given us silencers that'll work on a diverse variety of firearms. A .45-caliber suppressor will work on a 9mm pistol as well as a .45-caliber pistol. Pistol silencers will work on pistol-caliber carbines and in some cases with subsonic 300 Blackout. Likewise, a .30-caliber silencer will work on smaller rifle calibers such as 5.56 NATO. Most modern .22-caliber cans will silence 22LR, 22 Magnum, and even 5.7mm.

The SilencerCo Hybrid 46M is a true modular suppressed system that grants the shooter a variety of options to run on a multitude of firearms.

If you're running the SilencerCo 46M on a bunch of different firearms at a single range session, you might want to install the larger end cap in order to avoid a disaster while shooting.

Anderson's AR-15 with wood furniture courtesy of Boyds Gun Stocks proved to be a very quiet setup with the Hybrid 46M in full-length mode.

Yet finding one to take care of everything eluded us for years. A silencer for 45-70 would be too big to handle 22LR. Most rifle suppressors were too heavy to work on pistols, and there are enough completely non-standard thread pitches in use to make you rethink the whole idea.

SilencerCo, however, kept their nose to the grindstone and brought one out late last year called the Hybrid 46M. It was an evolution of their earlier Hybrid 36 and 46 designs. The key difference on this one and the Hybrid 46 is its modularity. You can shed about 2 inches in length and a little over 2 ounces in weight by running the shorter version.

The SilencerCo Hybrid 46M will work on just about any firearm with a threaded barrel so long as the caliber of the projectile is under 0.46 inch. That means it'll work on 45 Colt, 45 ACP, 45-70 Government, 458 Socom, 338 Lapua Magnum, 308 Winchester, 5.56 NATO, 300 Blackout, 9mm, and so on.

A WORD ABOUT ENDCAPS

SilencerCo offers a few different endcaps for the Hybrid 46M. This is good in a sense because the smaller you keep that exit hole, the more gas you prevent from escaping and creating a sound signature. It may keep your shots quieter by a few decibels if you have a 0.308-inch endcap on your 7.62 rifle as opposed to a 0.45-inch endcap. If you choose to only run a few calibers, that's fine.

However, it can be easy to forget to swap out endcaps if you're running a variety of different guns with this can at a range session, and small parts like that have a habit of getting, shall we say, misplaced. While 5.56 and 7.62 can sail through a 0.45-inch endcap with no problem, the reverse isn't true.

Your ears may not be sensitive enough to detect a difference between 2 and 5 decibels, so it may be wise to leave the bigger endcap on there because an endcap strike causing damage to the can or to the shooter would be a very bad day indeed.

The Hybrid 46M is more than capable when run in its shorter size.

THE TEST GUNS

To try this on as diverse a variety of firearms as possible to show how effective the SilencerCo Hybrid 46M can be, we tested it with the following guns:

- A stainless Marlin Model 1894 in 357 Magnum
- An Anderson Arms AR in 5.56 NATO
- A Remington Model 700 in 7.62 NATO
- A Glock 19 in 9mm

For these firearms, the choice of endcap was easy — stick with a .35-caliber one to accommodate the

The Charlie piston system (sold separately) allows the shooter to install a booster system to run the 46M on a variety of handguns.

357 Magnum and 9mm. There was no reason to step up to a .45-caliber endcap and no need to step down to a smaller one for .308 or 5.56 NATO.

MARLIN 1894

This lever-action rifle is a particularly fun one to shoot. Sometimes, it has issues feeding 38 Special rounds through it, due to the length of that case compared to 357 Magnum. Which means, with factory ammo, the round is quite loud to begin with. The test rifle is equipped with a Skinner peep sight as a rear sight and the company's patented bear buster with a white face for improved visibility, which is handy when shooting with a can.

In this case because the SilencerCo is a true rifle silencer instead of a beefed-up pistol suppressor, the sound signature was impressively reduced even in short mode. This was to be a sign of good things to come with the Hybrid 46M. If it could tame this beast, it should deal well with the others.

ANDERSON ARMS

ARs are typically pretty easy to suppress but can often provide excessive blowback from being over gassed. The SilencerCo Gas Defeating Charging Handle solves that problem, as it was designed with suppressed shooting in mind. It accomplishes this by using a rubber O-ring on the underside and is machined for a tighter fit. It's completely ambidextrous and performs well.

This particular rifle is outfitted with laminated wood from Boyds Gun Stocks and a Lucid Optics red-dot sight. Even in short mode, the suppressor was completely hearing safe with 5.56 NATO, and there was no excessive gas courtesy of the SilencerCo charging handle. An ASR (Active Spring Retention) flash hider from SilencerCo made mounting easy. Once shouldered, a locking ring ensures the silencer won't rotate or work itself loose under a heavy firing schedule. And their resonance suppression effectively eliminates the tuning-fork effect common with three-prong flash hiders.

REMINGTON 700

One of the benchmarks for testing rifle suppressors is mounting it onto a bolt-action rifle like the venerable Remington 700. In this case, we chose an ASR muzzle brake. The reason for using a muzzle brake over a flash suppressor is that the construction of the brake acts as a sacrificial baffle to prolong the life of the suppressor, as the initial blast isn't directly hitting the baffles inside the can. The can sounded particularly good on this rifle in both long and short mode — so much so that you could easily only run it short.

The Remington 700 is one of the most popular bolt-action rifles in the USA, and the Hybrid 46M mounts up quickly and lets you shoot quietly.

This Marlin 1894 always seemed loud while running 9mm cans on it. The larger volume of the 46M and the fact that it's primarily a rifle silencer instead of a beefed-up pistol suppressor made it extremely quiet.

GLOCK 19

If this were your typical rifle suppressor, this would probably be the end of the story, but not in this case. There's a separate accessory available for the SilencerCo Hybrid 46M called the Charlie Piston Mount. This includes a spring-loaded Nielsen device necessary to run a semiautomatic pistol with the 46M as its silencer. Retail is about $129, and a piston with the correct thread pitch will set you back another $86.

You remove the rear cap, replace it with the Charlie mount, then screw it onto the pistol barrel of your choice. Our test host was a Glock 19 with a 13.5 Metric left-hand pitch.

The SilencerCo was slightly heavier than a run-of-the-mill 9mm can, but its performance in the short configuration was solid with 147-grain ammunition. It's amazing that one silencer can finally handle these different firearm types as well as calibers. We've still barely scratched the surface on the potential of the Hybrid 46M.

FINAL THOUGHTS

SilencerCo doesn't recommend using rimfire ammunition with the 46M, which is understandable as rimfire ammunition is particularly dirty. A few rounds once in a while couldn't hurt though and using it afterward on a high pressure rifle will probably blast out any accumulated lead fouling. In any case, dedicated 22 cans are affordable to keep on hand for that role.

For a single silencer to suppress almost every firearm in your safe (excluding shotguns, .50 BMG, and rimfire), it makes a lot of sense. This is a great platform for someone to start out with, especially if you have a diverse collection and don't look forward to numerous tax stamps and wait times. In that sense, the higher price tag of the 46M and its accessories are more justified, rather than filling the coffers of the ATF.

On a Glock 19, the Hybrid 46M was a great performer in short mode.

SILENCERCO HYBRID 46M
SILENCERCO

MODEL: Hybrid 46M
CALIBERS: 9mm to 45 ACP; 5.56 NATO to 45-70, 338 Lapua Magnum
WEIGHT: 12.2 to 14.9 ounces
LENGTH: 5.78 to 7.72 inches
DIAMETER: 1.57 inches
MATERIALS: 17-4 stainless steel, titanium, and Inconel
MSRP: $1,117
URL: silencerco.com

QUIET, COVERT, KILLER

We Make James Bond Proud with the Trailblazer LifeCard

By Dave Merrill

The LifeCard from Trailblazer Firearms doesn't appear to be anything close to a gun at first glance. But there's more to be seen here. While many would consider it a mere novelty (and there's nothing wrong with that), we think there's a place for the LifeCard for those who prefer a martini shaken, not stirred.

OPPOSITE OF STUPID

It started with a horrible television show. Now-imprisoned Will Hayden of Red Jacket Firearms made a horribly clumsy and awkward version of the prototype Magpul FMG9 (which in turn was inspired from other folding designs such as the ARES Stealth gun). To say Hayden's horrible design sparked inspiration would be technically correct, but not for the usual reasons.

Sitting in a hotel room, Aaron Voigt shook his head in plain disgust, deciding to design the smallest possible gun he could make that could still fire .22LR, while also not being federally regulated.

While pen guns and the like have been around for a long time, they fall under the purview of the federal government and require a tax stamp as an Any Other Weapon (AOW). AOW is a catchall for many items, and anything that shoots that doesn't immediately appear to be a firearm falls under this category.

Voigt doesn't have a background in engineering. He was a construc-

tion worker by trade since the day he left high school, only detouring for a short stint in the military before returning. He even founded his own company, but when the 2008 housing crisis hit, he shuttered his doors and became a catastrophic insurance adjuster. This meant long days on the road, and a whole lot of awful television in hotel rooms. We're sure being victim to horrible hotels is the only reason anyone watched that show in the first place.

After watching that abortion of an episode, Voigt took the hotel card key from his pocket and traced it on a piece of paper. This would be the baseline size, and indeed what ultimately became the LifeCard.

Voigt sketched and designed for more than a year. He built a rough draft from wood, and put it into SolidWorks. By the end of 2011 he had one functioning prototype, which he promptly put in his pocket and brought to SHOT Show in January, 2012. Almost immediately, he had a deal with a current manufacturer, which fell through due to one of many panics where nothing aside from AR-15s were made. Another deal was struck with another company, which led to a couple more years of development.

And that didn't work out either — so Voigt said "screw it," and decided to do it himself in early 2015. After nine months of waiting

The GSL Pill Box is deceptively simple.

on the BATFE to ensure it wouldn't be considered an AOW, plans for construction could finally begin. Tooling finally hit metal in 2017, and he's been churning out the LifeCard ever since.

FEATURES AND QUIRKS

The LifeCard .22LR is a single-shot, folding pistol. One latch allows the entire unit to unfold into something resembling a pistol, and another latch allows the barrel to tip up to load.

While the original functioning prototype included a folding blade, that feature wouldn't be on production models, instead there's a small storage container for three .22LR rounds.

If you think unfolding a gun would be horrible for a self-defense handgun, let alone the fact it's a single-shot .22LR, you'd be absolutely correct. However, the LifeCard was a pistol designed with size and concealment in mind over every-thing else. We're as far as could be from advocating a .22LR, but the covert and murdery nature of the Trailblazer LifeCard really gets our hearts moving.

After the pistol is unfolded (it can be safely stored loaded), the hammer has to be cocked before the shot can be fired. After firing, you reload by hitting the barrel latch and manually extracting the spent casing before inserting a fresh round. This definitely isn't a speedy affair.

In order to re-fold the LifeCard after firing, the hammer has to be pulled back slightly to reset the trigger. We really appreciated that the LifeCard shipped with a snap-cap. Dry firing rimfire .22LR guns is particularly destructive, and the bright orange snap-cap allows for it without damaging the chamber face (you should see how mucked up our snap-cap got from dry firing).

The sights are a mere reminder that guns are supposed to have them, consisting of a shallow trench. But let's be real, this isn't meant for the target range. Accuracy is about what you'd expect, but accuracy isn't what you need if you're pointing it at contact range behind someone's ear.

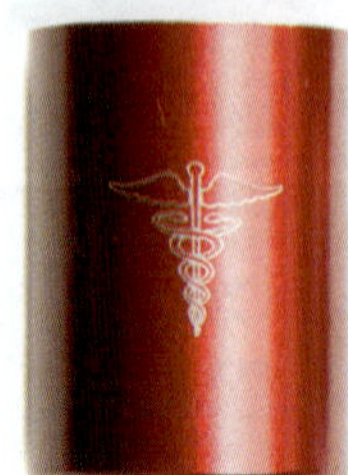

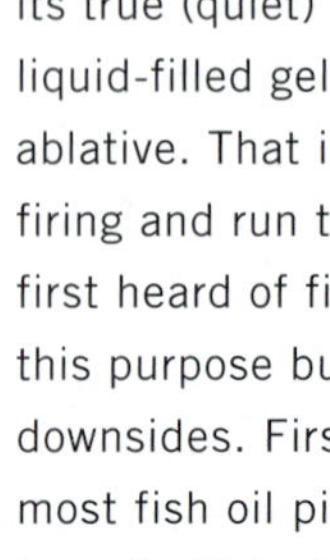

BOND'S PILL BOTTLE

A while back we acquired a GSL Pill Box, a current version of the original clandestine .22LR Gemtech Pill Bottle. Can we call it a "version" if the original was made by Greg Latke of GSL? We digress; the GSL Pill Box is a simple, ingenious design. It consists of a single chamber with a single thick wipe to contain the blast.

Disguised as a pill bottle you carry on a keychain, a false top is removed to reveal ½x28mm threads. The normal cap is removable so it can behave as a regular pill bottle.

In fact, holding pills inside is an advantage — not only does it hide its true (quiet) nature, but some liquid-filled gel caplets act as an ablative. That is to say, burst upon firing and run the Pill Box wet. We first heard of fish oil pills used for this purpose but there are two downsides. First and foremost, most fish oil pills are simply too large to fit inside. When we purchased smaller fish oil pills for the same purpose, the entire silencer, guns, and our hands smelled like bad tuna fish for days.

Instead of fish oil pills, Gas-X pills are incredibly small, very inexpensive, and don't make you smell like a cat food factory. Plus, we get a snicker out of using Gas-X pills to help quash the noise of gas expansion.

The traditional host for this covert can would be a Beretta 21A Bobcat — but after watching too many James Bond films (Craig was the best, Brosnan was the worst), we wanted smaller.

The LifeCard caught our eye originally as a curiosity, but we realized it would be the perfect host for our Pill Box. The only trouble: There's exactly zero exposed barrel to thread — so there was some engineering to do.

ADAPT AWAY

We got on the horn with Voigt and told him of our idea and plan. To get around the lack of a threaded barrel, Voigt decided internal threading in a proprietary pitch could be used to fit an external threaded adapter. This has been done to many firearms over the years, just never to a Trailblazer LifeCard. Soon enough, we received a revised LifeCard.

Not everything was instantly awesome, though. The thread adapter provided by Trailblazer didn't have any flats — meaning we were just as likely to unthread the adapter as we were the silencer. While in most instances we'd simply apply some Rocksett to make this problem disappear, we wanted to be able to easily remove the adapter for concealment purposes.

After some minor Dremel time, we had rudimentary flats on either side of the adapter, allowing for easy removal. A touch of black

Voigt's first wooden mockup.

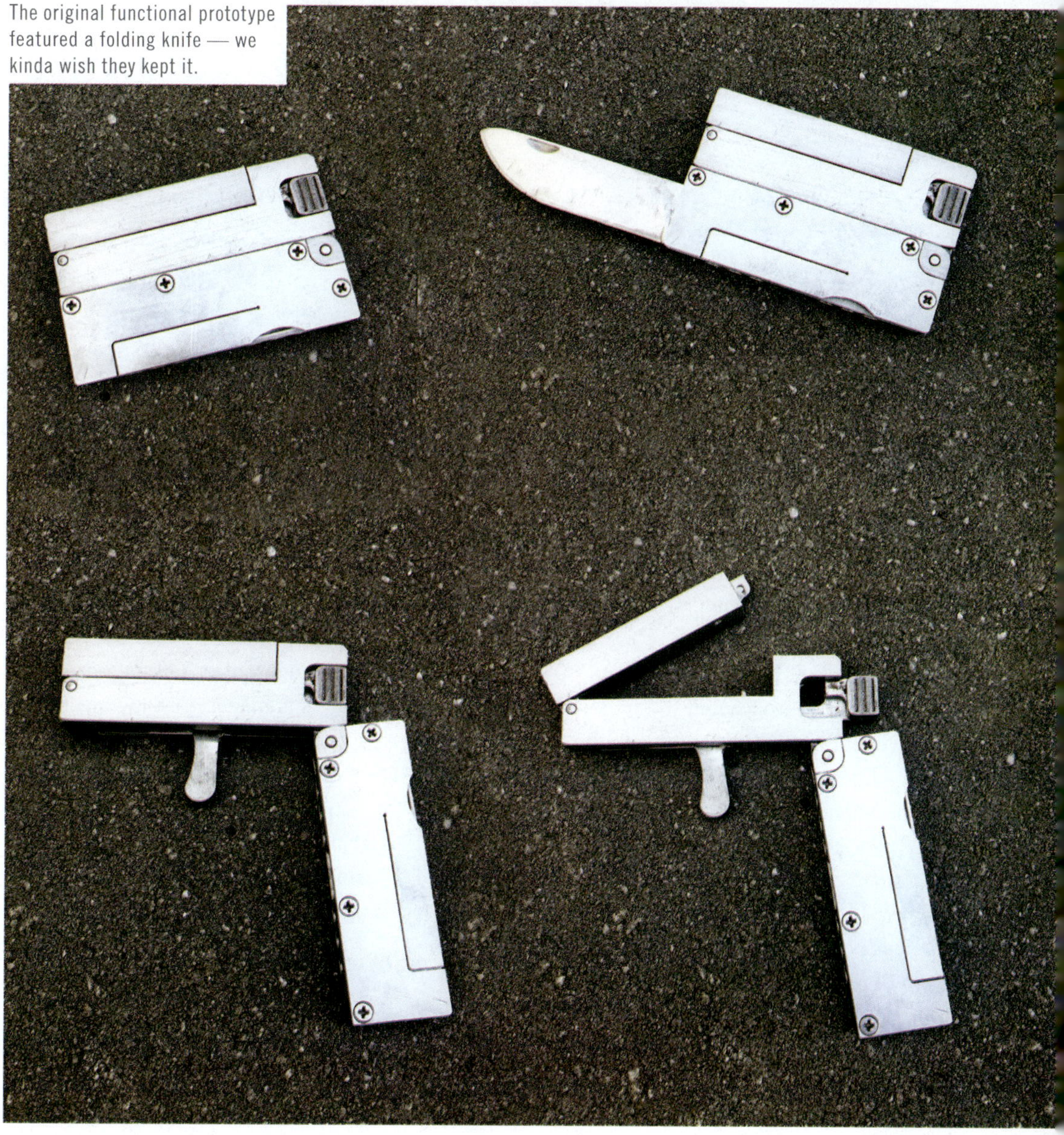

The original functional prototype featured a folding knife — we kinda wish they kept it.

Q from James Bond, eat your heart out.

paint and it was like no one knew they weren't there in the first place.

HIDDEN HEATERS

So of course, we wanted to discover and engineer all sorts of ways to hide this gun and silencer combination. The silencer was the easy part — it's already disguised as a functional keychain pill bottle. But the firearm and adapter took more work.

The LifeCard itself fits in an empty pack of cigarettes. However, smokes don't weigh that much and don't normally ding on metal detectors. This will get the LifeCard into some, but not all, places.

Similarly, one could simply embed the LifeCard inside the sole of a steel-toed boot and not have any problems unless an X-ray machine was present.

We decided to go full Q from James Bond and engineer something from scratch. First, we gutted a portable battery pack and made a specific compartment for the LifeCard and thread adapter. Then, we went the extra mile to ensure the battery pack would still work by rewiring it to a single cell.

It's still a functional external battery pack but with a surprise inside. Though it won't evade X-ray machines (we have a version that does that too but won't share it, because f*ck terrorists), it does provide plenty effective camouflage for most situations.

Given the small size, and the fact that it's a not instantly recognizable as a firearm, you could come up with a dozen different ways to carry this gun in places where you couldn't otherwise carry a gun.

LOOSE ROUNDS

With regard to practicality? No, the LifeCard shouldn't be the only pistol you ever own. And sometimes

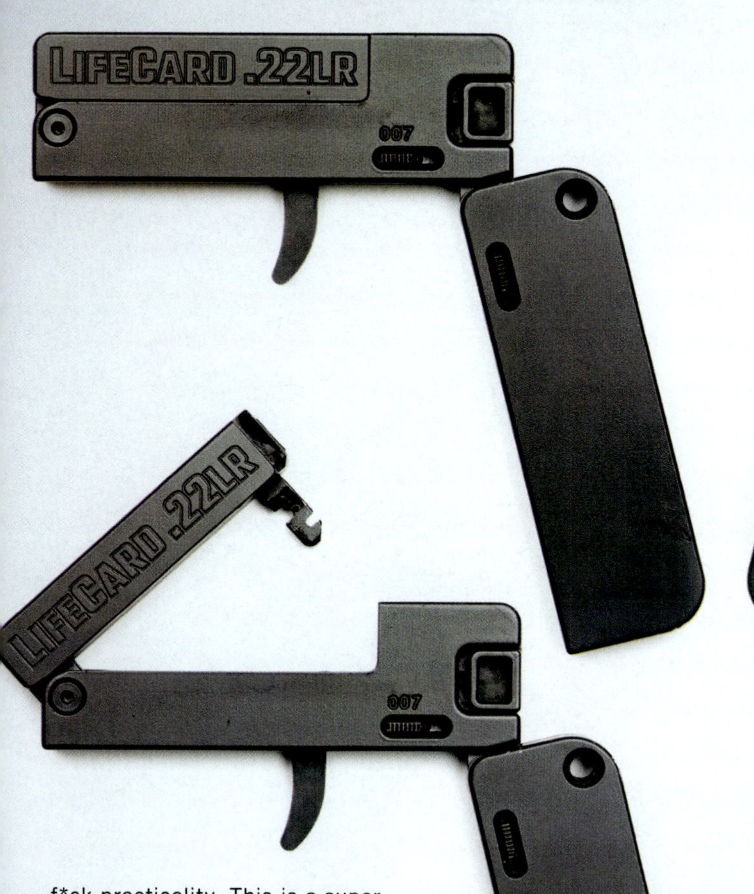

— f*ck practicality. This is a super cool little covert killer that's beyond fun to show off to your friends and at the range. With a thread adapter, and in combination with a stubby silencer such as a GSL Pill Box or a Bowers Bitty, it becomes a quiet little dispatching tool.

We've heard some RUMINT that we'll all soon be seeing some different models, and we're holding our breath.

This year at the 2019 NRA Annual Meeting, Trailblazer was showing off a threaded version that'll hopefully be available to the general public soon. You're welcome.

TRAIL[illegible] FIREA[illegible] LIFECARD .22LR

CALIBER: .22 LR

WEIGHT UNLOADED: 7oz

MAGAZINE CAPACITY: 1 round

LENGTH: 3.375 inches

WIDTH: 2.125 inches

THICKNESS: 0.5 inch

MSRP: $399

ACCESSORIES:
GSL Pill Box + $285
Threaded Barrel Adapter + $TBD

AS TESTED: $684+

URL: Trailblazerfirearms.com

IS THAT A SILENCER IN YOUR POCKET?

Cans Small Enough to Carry

By Alexander Crown

More than 2.4-million silencers have been registered in the last 10 years — there's no denying the proliferation of suppressors in today's market. Consumers have more choices now than ever before at almost all price points, wide availability, and with unique features to suit their needs. Here we'll examine a few uncommon choices for pistol suppressors that are filling a segment of the market.

For many years now, modular, multicaliber suppressors have been extremely popular. This makes sense, as not only do suppressors carry a price tag relatively similar to another firearm, but you also have to endure a $200 tax as well as a lofty wait time for processing of the Form 4. These hurdles have led consumers to purchase cans that can do almost everything, which is all well and good but can lead to underperformance compared to a purpose-built, caliber-specific item. Many suppressor owners come to find this out and then set their sights on models that are designed for a specific caliber or other parameters.

Here, we'll examine options that are small, as in they can fit in your pocket. These suppressors all share one common feature, or lack thereof: they don't require a piston to function.

Commonly called a Neilson device, booster, or piston system, this is the mechanism that relieves or decouples the weight of the suppressor off of tilt-barrel pistols (most pistols produced today) during cycling to allow them to function properly. Generally, hanging a weight off the end of that barrel will cause malfunctions. So, suppressor manufacturers use a Nielson device to counter this. Piston systems add to silencer length, and so long as a suppressor is sufficiently light and your ammunition appropriately strong, you may not need one.

Some of these models are modular and require a booster in longer/heavier configurations but can run without them when shorter. According to Blake Young, chief engineer for Primary Weapon Systems, holder of nine suppressor related patents, and designer of two of the cans shown here, the magic number is four — roughly 4 ounces. If a suppressor weighs or can be configured to weigh 4 ounces or less, it's far more likely to work without the piston system.

It's also important to note that in order to get even moderately decent sound performance from these little cans they require wipes and/or being made wet. For more information on making cans quieter, check out RECOIL Issue 58, where we go in depth on tactics to increase sound performance.

Lastly, all of the mentioned suppressors are designed for 9mm but can be used on lesser calibers like .32 ACP, .22LR, .25ACP, and others.

THE CANS

GEMTECH AURORA 2

The Aurora 2 is an updated version of a silencer produced in the mid '90s by Gemtech. This suppressor uses old technology in the form of wipes, spacers, and petroleum jelly. By stacking the wipes and spacing them with hollow spacers, the Aurora 2 creates a channel for the bullet to travel through. All wipes and spacers are coated with petroleum jelly to assist in absorbing the hot gases. This creates an artificial environment that temporarily lasts for several rounds. The other unique feature of the Aurora 2 is that it's a dual-threaded can, meaning that one end is threaded standard ½-28 and the other is 13.5x1mm left hand, so you can use it on practically any 9mm handgun with a threaded barrel without the need for adapters.

Diameter: 1.125 inches
Overall Length: 3.5 inches including thread protectors
Weight: 3.5 ounces
Mount Type: ½-28 or M13.5x1 LH thread at muzzle
Material: 7075-T6 aluminum
Finish: MIL-A-8625 type II, class 2 black anodized
MSRP: $399
URL: gemtech.com

THOMPSON MACHINE POSEIDON 9

The Thompson Machine Poseidon 9 also uses a wipe and petroleum jelly to create an artificial environment. The Poseidon differs from many in that it utilizes a small mono-core, making it easier to disassemble, clean, and relubricate after firing. The mount portion of the can has interchangeable thread adapters for use on ½-28, 13.5x1mm LH, and ½-36.

Diameter: 1.25 inches
Overall Length: 4.125 inches
Weight: 3.8 ounces
Mount Type: ½-28, ½-36, 13.5x1mm LH
Material: 6061T6 aluminum
Design: Monocore, take-apart tool included
Finish: Anodized matte black
MSRP: $379
URL: thompsonmachine.net

JK ARMAMENT JK 105 CCX 9MM

Originally available as a so-called solvent trap, the JK 105 CCX is roughly the size of a modular .22LR silencer. Like all other JK Armament suppressors currently available, the CCX is modular, with configurations as short as 2.37 inches up to 6.25 inches. In the smallest setup, the JK 105 CCX weighs in at a mere 1.5 ounces. This model doesn't use any wipes but an ablative like petroleum jelly is highly recommended for increased performance. Though ostensibly for 9mm (hence the name), other compatible calibers are .380ACP, .32ACP, .30SC, .38 Super, and .38 Special.

Diameter: 1.05 inches
Overall Length: 4.13 inches
Weight: 2.3 ounces (without mount)
Mount Type: ½-28
Material: 7075 aluminum
Finish: Type 3 hard coat anodized
MSRP: $405
URL: jkarmament.com

THUNDER BEAST ARMS FLY 9

The Thunder Beast Fly 9 is a full-size suppressor but can be taken down to a smaller size. We're told that when in the small configuration and without the added weight of a piston system, it can operate on many handguns. While performance will be drastically degraded, it can be made wet to improve sound reduction. This can is very lightweight, even in the full-size configuration, but taking it down makes it fit into our category of pocket cans.

Diameter: 1.39 inches
Overall Length: 7 or 4.4 inches
Weight: 7.9 ounces (full size); 5.5 ounces (short), both with the piston system
Mount Type: ½-28
Material: Titanium
Finish: Cerakote black, OD green, FDE
MSRP: $1,095
URL: thunderbeastarms.com

LONE WOLF ELIMIN8R

Another honorable mention is the Lone Wolf Elimin8r. We covered this suppressor in RECOIL Issue 57. Although this is a .45 ACP suppressor, the Elmin8r uses eight individual baffles, allowing the user to choose how many baffles they want to utilize. If the main body plus one baffle and the endcap are used, the suppressor will weigh just under 4 ounces, aiding its ability to be used without a booster. This can also features O-rings that help it come apart after firing so the user can swap it easily to whatever size they desire.

Diameter: 1.375 inches
Overall Length: 8.45 inches
Weight: 10.1 ounces
Mount Type: ½-28 or .578-28
Material: 7075 aluminum
Finish: Black anodized
MSRP: $600
URL: lonewolfdist.com

REX SILENTIUM SEG H GEN2

This is another modular suppressor that allows users to choose their desired size. With the main body plus one baffle and endcap, the can weighs 3.3 ounces; adding one more baffle increases it to 3.9 ounces, allowing it to function with only a thread mount. Rex does recommend using both the included wipe and an ablative to maximize sound reduction when using the suppressor in the small configuration.

Diameter: 1.37 inches
Overall Length: 8.6 inches
Weight: 11 ounces
Mount Type: ½-28
Material: 7075 aluminum
Finish: Black anodized
MSRP: $600
URL: rexsilentium.com

LOOSE ROUNDS

Each of these suppressors have different features and capabilities that make them a worthy choice for your specific needs. With the huge surge in increased capacity and everyday carry handguns, users may be on the hunt for a small suppressor to keep in their bag for various reasons. You couldn't go wrong with any of these offerings.

PWS BDE SUPPRESSOR

Primary Weapons Systems Comes Out Swinging with its First Suppressor

By Steven Kuo

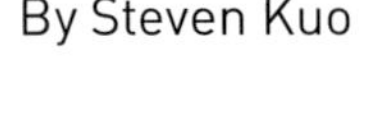

The PWS BDE is 3D-printed titanium, with four removable baffles so you can configure it to the exact mix of length and performance that you desire.

Primary Weapons Systems is best known for its line of piston-driven AR guns, jumping early into the arena of expanding the technological envelope of the mainstream AR platform. They've now turned their sights on the fast-growing suppressor market, releasing their first-ever silencer, the BDE Suppressor. PWS says the moniker stands for Bravo Delta Echo, but a quick Google search may provide an alternate yet even more appropriate meaning.

The BDE is composed of titanium that's 3D printed and CNC turned, allowing for precise and intricate designs, concentricity, and use of a strong and lightweight material. It's anodized and Cerakoted, with exterior texturing designed to help dissipate heat mirage and look snazzy in the process.

It's modular, with four removable baffles to exactly configure the BDE to your desired length and performance. The shortest configuration with just the end-cap is 6.5 inches; screwing on all four provided baffles increases the length to 8.2 inches. Overall weight ranges from 12.7 to 17.4 ounces. Those who prefer fewer moving parts might wish that PWS had provided two longer double-baffled segments rather than four singlets, but the BDE features tapered threading to help prevent binding or carbon locking while sealing gases and preventing loosening. Additionally, the internal baffle notches are symmetrical to optimize accuracy and the serialization is on the blast chamber, so you can easily upgrade or service it in the future.

As you'd expect, removing baffles sacrifices noise suppression performance. PWS' test results range from 121.5 dB in the longest configuration at the shooter's left ear with an 8-inch 300BLK bolt gun to 140.7 dB in the shortest with a 16-inch .308 direct-impingement AR.

The BDE ships with a 5/8-24 direct-thread adapter, but you can easily remove it and attach any 1.375x24 HUB mount adapter that you'd like to the entrance chamber. PWS also supplies two wrenches; you can snap them on a 3/8-inch torque wrench to tighten baffles to spec. At 1.75 inches in diameter, the BDE's a bit girthy.

PWS rates the BDE for full-auto use, with no barrel length restrictions, and accommodates up to .300 Win Mag. We replaced the supplied direct-mount adapter with a Dead Air Silencers KeyMo HUB adapter and tested the can on the 11.5-inch Tinck Perun reviewed elsewhere in this issue. The BDE was comfortably hearing-safe and a bit quieter than some other suppressors in the stable. It was a delight to shoot, especially as the Perun is an extremely soft shooting gun. The Perun is also a gas-piston gun; we occasionally felt a bit of gas blowback, but most of the time we didn't notice any.

PWS has really come out swinging with their new BDE Suppressor, and we look forward to what they'll do in the future given its modularity. The BDE certainly put a spring in our step when we tested it. R

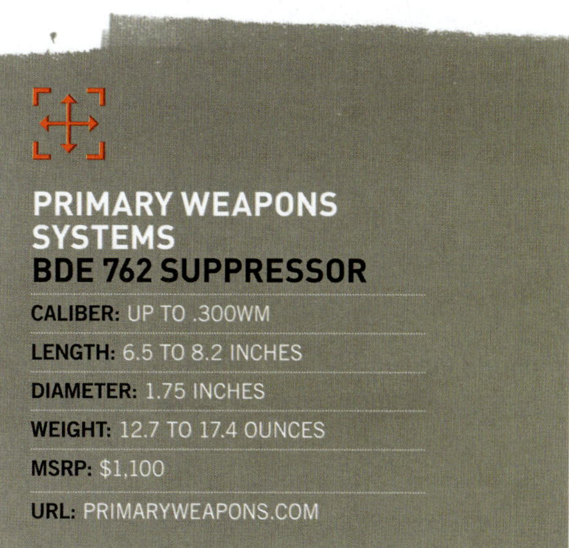

PRIMARY WEAPONS SYSTEMS
BDE 762 SUPPRESSOR

CALIBER: UP TO .300WM

LENGTH: 6.5 TO 8.2 INCHES

DIAMETER: 1.75 INCHES

WEIGHT: 12.7 TO 17.4 OUNCES

MSRP: $1,100

URL: PRIMARYWEAPONS.COM

ABEL SUPPRESSORS

Short, Chunky, Effective. The Biscuit is Equally at Home on SBRs and Bolt Guns

By Iain Harrison
Photos by Kenda Lenseigne

If someone is looking for their first can, we almost always steer them toward a .30 cal for maximum versatility. It seems the engineers at Abel are on the same page, as for the moment, that's all they offer — which is a good thing. We got to spend some quality time with their Biscuit and Biscuit S models and can confirm they're as good as anything out there, offering a lot of bang for the buck.

Constructed of 17-4 precipitation-hardened stainless steel and featuring some of the nicest welds we've seen on any suppressor, they're rated for use on 5.56 rifles down to 10.3 inches in barrel length and are claimed to withstand full-auto fire on shorties. At the other end of the spectrum, they'll handle 300PRC from a 22-inch bolt gun, and we'd be surprised if 300 Norma wasn't on the cards also. Both models are built around the industry-standard

1.375x24 HUB mount, meaning many other suppressor companies' adapters will bolt right up. So, if you're already invested in say, ASR or KeyMo brakes and flash hiders, they'll fit your guns without much hassle. The Biscuit measures 1.8 inches in diameter instead of the usual 1.5, which means it achieves the same internal volume as skinnier cans, but in a shorter overall length. You might think that being a little chunkier, it might be more noticeable when viewed through a scope, and we're glad to report that nope, the greater diameter is offset by the shorter length and it's no more obtrusive in your field of vision.

Of the two, our preference is for the Biscuit S. Yes, it's louder. No way around that. But when paired with a firearm with a shorter OAL, such as the Springfield Hellion shown here, it really shines. In fact, the Biscuit S might just be the ideal bullpup can, as it keeps length to a minimum, adds just enough weight to the muzzle to offset the usual bullpup butt-heavy feel and, due to its low back-pressure design, allows the gun to operate on the normal gas setting.

ABEL
BISCUIT, BISCUIT S

CALIBER: .30

LENGTH: 5.5, 4.7 inches

WEIGHT: 13.5, 16 ounces

MATERIAL: 17-4 PH stainless

MOUNTING SYSTEM: 1.375x24

URL: abelcousa.com

The Biscuit S pairs perfectly with the Hellion; it doesn't add too much length while also improving the balance.

SHOOTING FOR SILENCE

Five Lightweight Silencers Ideal For Hunting

By Dave Merrill

Forty-two of the 50 states in the USA allow you to legally hunt game with a silencer. There are a mere two states where suppressors are legal for general use sans hunting: Vermont and Connecticut. There's a weird exception in the state of Maine … because a special, separate permit must be issued to use a silencer for hunting — but that's it.

In fact, we expect the hold-out states who allow silencers for everything except hunting to perhaps sway our way in the coming months. Some actual common sense? Mayhap. In a few minor ways, we're finally reaching the rest of the developed world in the realm of suppressor use.

Frankly, there are a ton of damn good reasons to use a silencer for hunting. First and most obvious is the reduction of your sound signature — and this works both ways. Not only are you far less likely to have ringing ears post-shot, but nearby animals are less likely to have a reaction to the auditory explosion your trigger pull created.

Individual ownership of silencers has been steadily increasing for more than a decade now, and people usually aren't satisfied with just one (that's what she said?). The market has reacted to this, increasing the availability of silencers specifically designed for game-getting.

While any old suppressor should do something to reduce the noise from your muzzle, today we look at hunting-specific silencers.

The same aspects that make some suppressors balls-out amazing for hanging off the end of a 7.62N belt-fed FN 240 Golf machinegun make it f*cking stupid for a hunting rifle. For hunting we don't need one that withstands automatic high-pressure shots with steadfast durability; we need a silencer with less weight, decreased size, and good-enough capability.

Think about it: Dedicated hunting rifles prioritize weight-savings over almost anything else. They aren't made for a 250-round course of competitive fire at a PRS match, nor for being deployed as a

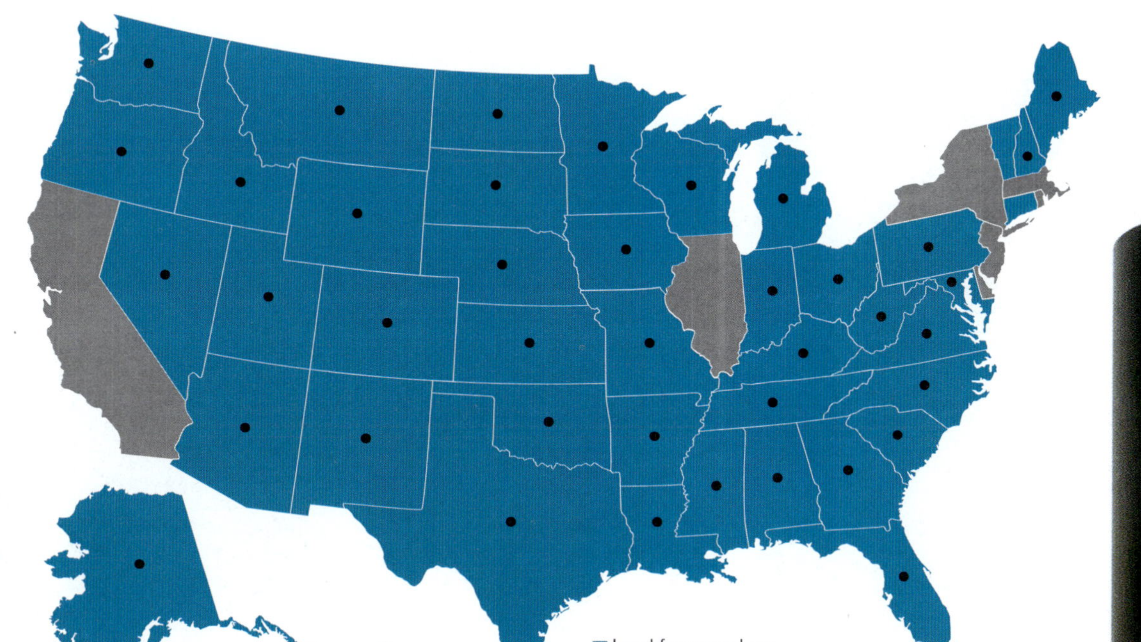

sniper rifle overseas. The hunting gear you see advertised purports the perfect balance between weight and capability.

We wish there were a silencer that was both featherweight and über-durable, but that simply isn't the case. Today, we look at silencers mostly designed with hunting in mind, discussing some of their idiosyncrasies. This is a selection of .30 caliber and above silencers, weighing no more than 16 ounces, most all of which are meant to live 24x7 on the end of a rifle. You'll also note that we don't list decibel reductions with any of these silencers. All of them are intended to be multi-caliber, and therefore will have varying results depending on the rifle, barrel length, and caliber used.

There are some standouts on the list, which we'll point out along the way. Take note that all specifications are based on our own measurements, as some silencer companies have been prone to list weights in a non-shootable configuration, such as without required thread mounts or muzzle devices.

GEMTECH **TRACKER**

CALIBER:
.300 Win Mag & calibers below

LENGTH:
8 inches

WEIGHT:
11.1 ounces

MATERIALS:
Aluminum

MSRP:
$599

URL:
gemtech.com

NOTE:
Note that the Gemtech Tracker is the only silencer we looked at for this buyer's guide that was actually lighter than the manufacturer-specified weight. Maybe it's because it only comes with a 5/8x24 direct-thread mount so there's little wiggle room, but it still seemed significant enough to mention.

The Gemtech Tracker was initially developed for a specific foreign special-forces sniper cadre and then only later adopted as an American hunting silencer. Primarily constructed out of aluminum, the Tracker definitely won't withstand continuous rounds — you'll need a 24-inch barrel or longer with .300 Win Mag or at least 16 inches with 6.5 Creedmoor or .308.

GRIFFIN ARMAMENT
SPORTSMAN ULTRA LIGHT 338

CALIBER:
.338 Lapua & calibers below

LENGTH:
8.75 inches

WEIGHT:
14.5 ounces

MATERIALS:
Aluminum / stainless steel

MSRP:
$995

URL:
griffinarmament.com

NOTE:
Griffin Armament previously released a similar silencer, but it was only rated for .300 Win Mag. And though it was a mere 11.3 ounces, we couldn't possibly pass up a model just 3 ounces heavier that now chomps up 338 Lapua, provided the barrel is 24 inches or longer. Griffin tells us their new patented hybrid HEDP baffles deliver the durability of titanium baffles, but with the additional weight savings of aluminum.

While Griffin Armament sells a direct thread adapter, the Sportsman Ultra Light 338 will also fit on standard Griffin mounts.

LIBERTY SUPPRESSORS
SOVEREIGN

CALIBER:
.300 Win Mag & calibers below

LENGTH:
7.125 inches

WEIGHT:
15.8 ounces

MATERIALS:
Titanium / stainless steel

MSRP:
$799

URL:
libertycans.net

NOTE:
While not initially intended to be a dedicated hunting silencer, the Liberty Sovereign meets all of our requirements. Instead of a set thread pattern, the Sovereign ships with included interchangeable inserts for ½x28mm, 5/8x24, and a Liberty muzzle device. That covers several bases right off the bat. As an added bonus, Liberty Suppressors throws in an Armageddon Gear silencer cover to help cut down on any heat mirage.

Liberty tells us that the minimum barrel requirements are 18 inches or longer with .300 Win Mag or 14 inches or longer with .308.

THUNDER BEAST **ULTRA 7**

CALIBER:
.300 Rem Ultra Mag & calibers below

LENGTH:
7 inches

WEIGHT:
11.4 ounces

MATERIALS:
Titanium / stainless steel

MSRP:
$1,045

URL:
thunderbeastarms.com

NOTE:
The TBAC Ultra 7 is the Goldilocks of TBAC's .30-caliber rifle can lineup. Its siblings are 2 inches shorter and longer, leaving the middle child long enough to be hearing safe, but short enough to have minimal effect on the handling of a typical 22- or 24-inch hunting barrel. TBAC's helical baffle stack and precise, 360-degree welding translates into accuracy the brand is renowned for.

Speaking of swapping rifles, the Ultra 7 is rated for .30-caliber cartridges up to .300 Remington Ultra Mag on a 20-inch barrel and performs admirably as a sub-caliber can.

Q **HALF-NELSON**

CALIBER:
.308 & calibers below

LENGTH:
6.8 inches

WEIGHT:
12.3 ounces

MATERIALS:
Titanium

MSRP:
$849

URL:
LiveQorDie.com

NOTE:
Q's Half Nelson fits more internal volume in the same space by upping the can's diameter from the industry standard 1.5 inches to 1.75 inches. This little tweak imbues the Half Nelson with more pressure-reducing volume than competing cans of similar length. Aside from increased sound mitigation, the larger volume translates into lower backpressure, good for gas guns, and lower temperatures, which helps with mirage.

The can was designed to be hearing-safe shooting .308 Win from a 16-inch barrel. The bore is made using a wire EDM process for accuracy, and Q says it's actually a tiny bit larger than competitor's .308 bores to reduce sound at the shooter's ear.

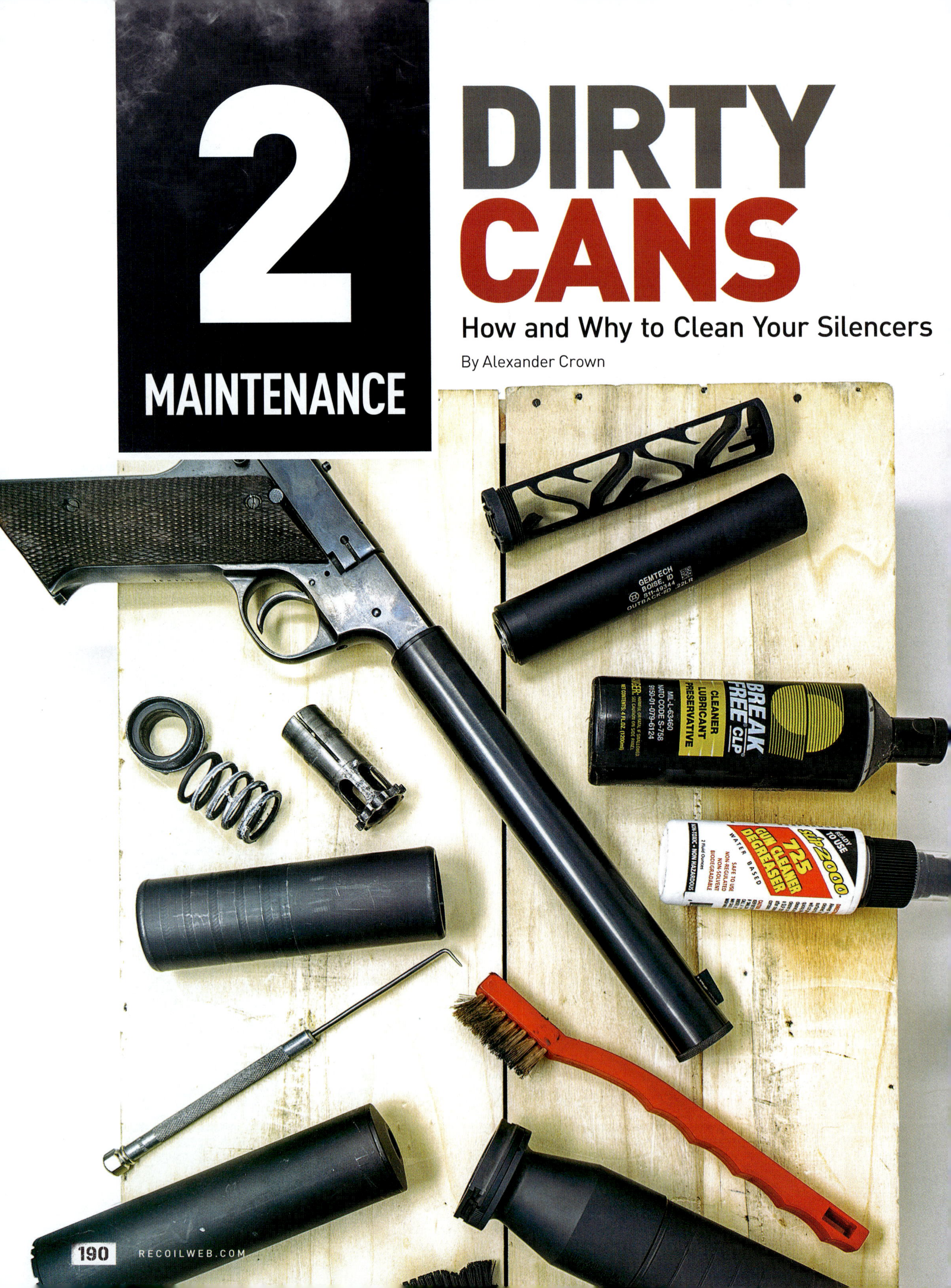

2 MAINTENANCE

DIRTY CANS

How and Why to Clean Your Silencers

By Alexander Crown

Giving your muzzle device a scrub helps prevent silencers from sticking.

We all know that guy who cleans his gun after every range trip, and we also know that guy who has never cleaned his gun. Neither are right or wrong but one thing we should all agree on is cleaning your silencer is unneeded, right? Well, sort of. Here we dive into some best practices about silencer maintenance, dos and don'ts, and the reasons behind them. Silencers are becoming very common place in today's firearms industry and rightfully so. Everyone should want less sound, improved ballistics, and to look cooler. The market is larger than ever before, and the number of cans is staggering. So, once you pay your $200 and wait your four to 15 months, how do you keep that silencer in tip-top shape?

The best way to break down silencer maintenance is to put them into categories: centerfire rifle, centerfire pistol, rimfire, and specialty silencers. Each of these types work differently, have different mounting systems, generally they have different materials, and ultimately, they're all maintained differently by the end user. The materials you'll come across most often are 6061 aluminum, 7075 aluminum, 17-4 stainless steel, and 6AL4V titanium. There are other metals being used and numerous coatings, but these are the most common. Stainless and titanium are pretty tough, allowing you to be a little more aggressive with your cleaning techniques where aluminum requires a gentler touch to prevent damage.

Booster assemblies should be cleaned and greased regularly.

RIFLE SUPPRESSORS

The centerfire rifle category of silencers is the easiest to deal with, as they require practically no maintenance. Some newer models come apart but if you look at these parts, they'll have little in the way of buildup of carbon, copper, or general gun funk. This is mainly due to the high pressure and extreme temperatures these silencers endure. The buildup inside burns off. Yes, the baffles will be discolored and dirty looking, but they don't need to be scrubbed to a sparkling clean, like they're new. Scraping the baffles may actually weaken them in some cases. The one area of rifle suppressors that should be cleaned is the mounting surface.

PISTOL SILENCERS

Centerfire pistol silencers also generally require very little maintenance. The most laborious part of cleaning the can will concern the Nielson Device, which is a "moving" part of the silencer. For the uninitiated, the Nielson device is the piston system that allows most semi-automatic handguns to cycle properly with the added weight of the silencer.

A spring compresses, essentially relieving the weight off the end of the barrel, allowing the handgun to complete its action. These devices require lubrication and cleaning. The piston assembly is generally made of steel and can be scrubbed to remove carbon buildup. After cleaning and inspection, a coat of white lithium or similar grease should be applied. This is recommended at different round counts from each manufacturer, but our rule of thumb is every 200 to 400 rounds.

Pistons that go unmaintained will become seized with the spring compressed. These can usually be removed and heated and/or pounded loose then cleaned. The rear portion of the silencer, where any mounts are installed, should also be cleaned of any buildup to maintain alignment. As for baffles in centerfire pistol silencers, unless very high round counts are involved or incredibly dirty ammo is used, they require little in the way of cleaning.

RIMFIRE CANS

Rimfire silencers do require regular maintenance. The rimfire cartridge is a nasty little bugger that gums up everything. At minimum, you should take your silencer apart after a range session and knock the debris out. This will help the silencer from becoming totally seized and unable to get it apart. Remov-

The owner of this poor silencer decided the best way to break carbon loose was with a ball-peen hammer. Do not do this.

Many silencers are user-serviceable, allowing you access to the innards.

ing baffles will be easier. A truly proven method to cleaning rimfire baffles is to scrub them with a stiff brush and CLP. Each baffle and the tube should be cleaned to the best of your ability. Most rimfire suppressors are thread-on models, but it's still important to check the mounting area for debris and keep it clean.

22LR silencers require extra care due to their easy accumulation of heavy metals.

SPECIALTY MUFFLERS

Specialty silencers include integrals and shotgun cans. Integrals come in all shapes and sizes, so knowing the materials used in their construction is a benefit, as this allows you to use cleaning products tailored for those specific metals. Shotgun silencers usually come apart and have aluminum and steel components, so treating them like a less dirty rimfire silencer will be best practice, if they need to be cleaned at all.

CLEANING TECHNIQUES

We're advocates of the slow process of scrubbing with a brush and CLP. This isn't a debate over cleaning products — when that happens everyone walks away dumber. Pick one you like and take your time with it; just be sure it's not caustic to whatever metal you have. Many Internet experts will recommend a process simply referred to as "The Dip."

Tumblers with the right medium can make cleaning some cans a snap.

The Dip is a mixture of chemicals that'll clean a silencer like nobody's business. The downside is that what you're left with is lead acetate. Lead acetate is extremely unsafe and can kill your children, partner, and pets — which is why it requires HAZMAT disposal. If that isn't reason enough, lead acetate will destroy aluminum.

Ultrasonic cleaners can be useful, where carbon isn't cooked on to the point of no return. Media tumblers can also assist in the cleaning of parts; however, using the correct media and amount of time is important to not have the finish of your silencer wear off. R

GETTING NEW GUTS

New Life for Your Old Silencer

By Dave Merrill

In the United States, silencers are somewhat special even among NFA items. It's a consumable item that'll eventually be worn out with enough use. And even if that silencer is kept in good condition, it still degrades in relative performance as newer, quieter, lighter, and better cans are released.

Combine all of the above with the fact that any individual NFA sales

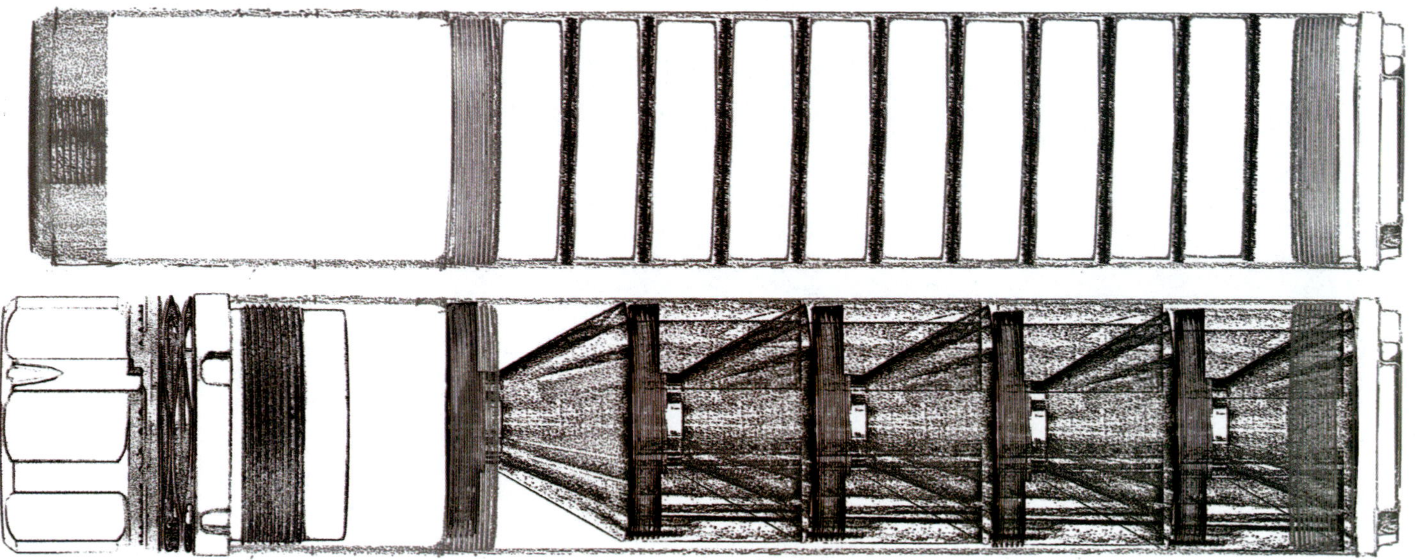

that are out of state require two tax stamps (in-state is only one) and you begin to understand why the secondary market for silencers is virtually non-existent.

A silencer is a lifetime purchase, but also an item that wears out. Awesome. What can we do? If a company is still around, even odds they'll service your old suppressor, but even that isn't guaranteed. Even if the company still exists, it doesn't mean that they have the current-best baffle stack or have the mounting system you want.

The old advice about silencers is to, "just go with a major brand, because they should have some longevity," but the 2016 post-election slump saw the beginning of the end of firearms companies of all types. AAC was one of the latest victims, following other silencer manufacturers such as Crux, Leviathan, Huntertown Arms, and more.

We've seen several new suppressor companies put up their shingles the last several years. And while we applaud the expansion of options for consumers and the innovation that arises from competition, it stands to reason that not all of these companies will be around in a decade or more.

NEW COTTAGE INDUSTRIES

When you buy a silencer and your tax stamp is approved, you now legally own and possess it. But, due to intentional obfuscation of laws, even though it's your own personal property, it can't be modified at-will. Silencers and psilocybin spores have more similarities than you thought. In order to modify or repair a silencer you have to be a licensed FFL allowed to work on NFA items. Set aside how ridiculous this notion is for a minute and you begin to grasp how large of a problem this really is.

There are several companies that'll perform specific re-coring jobs, often limiting themselves to a limited number of silencer models. This makes sense as it's not as if every manufacturer uses the same diameters and thread pitches; this is a job best left done to a custom shop. But there are some shops that offer more than modifying specific models. Today, we focus on Curtis Tactical.

We first featured their Modern Day DeLisle, the integrally suppressed Remington 700 chambered in 9mm that eats from Glock magazines in RECOIL Issue 51. While Curtis Tactical has their own line of suppressors, including several separate integral guns, their original bread and butter was re-coring older silencers.

To this day, Curtis Tactical dedicates a full 25 percent of their efforts to upgrading old silencers. Three times a year they spend an entire month replacing and upgrading cores. The process is both simple and somewhat tedious.

More than simple re-cores, the folks at Curtis Tactical can jailbreak your sealed silencer to make it user-serviceable and also replace or convert entire mounting systems. While we're sure Curtis Tactical would like you to change to their mounting system, importantly they

The Black Aces Po' Boy as it ships features flat baffles and spacers, but Curtis Tactical can not only convert it to a modern stack, they can change the whole mounting system.

It seems like every American machine shop has a line of manual machines in the back, and we love that these old manuals live alongside modern CNCs.

can convert silencers with legacy mounting systems to the now-universal 1.375x24 TPI HUB mount.

There are literally dozens of companies making silencers, mounts, and devices for the HUB mount. Having a hard time finding a given mount or have the old 51T AAC mount fail? Curtis Tactical can fix that for you.

Want a direct-mount left-hand Area419-compatible mount for precision work? They'll do that too. If you find factory service is no more — there's still hope.

MAKING CHEAP CANS GREAT

Because of the virtually lifetime purchase of a silencer, many view buying beer-money silencers as a false economy. In RECOIL Issue 46, we generally showed off a lot of budget options, and some silencers were among them. The Black Aces Tactical Po' Boy silencers sell for less than the tax stamp at $199 a pop. As we showed in that article, they don't feature the most modern of designs, largely consisting of flat baffles and rudimentary spacers. Of course, Curtis Tactical will pop a new core in those too, and the performance will unquestionably greatly improve.

Curtis Tactical has not just vestigial baffles in a box, but also pieces of history. Have a close look and you'll see more than a dozen companies represented, and some designs that possibly pre-date your own birth.

The hardest part about NFA paperwork is your first submission, because after you realize exactly how easy the entire process is, you just end up buying a lot more. The scariness disappears with experience. If a $199 silencer got your foot in the door, you're not stuck with those flat baffles forever.

And here's something really worth mentioning: Curtis Tactical will keep your original cores on-hand for a month. If you don't think their baffles are an improvement, they'll refund your money and put your old core right back in. They tell us that so far no one has taken them up on this offer, and we saw the vestigial guts of dozens of recent re-cores.

LOOSE ROUNDS

So long as silencers remain on the NFA registry, you're basically making a lifetime purchase. If the company you purchased from goes the way of AAC, places like Curtis Tactical can ensure you never run out of the mounting solutions and also keep that can full of new guts.

MAKING CANS QUIETER

Tips and Tricks to Make Your Silencers More Stealthy

Story by Alexander Crown
Photos by Alexander Crown and Dave Merrill

Suppressors, often called silencers (don't listen to idiots on the internet — both terms are correct), are a fantastic firearm accessory and can make shooting a significantly more enjoyable experience. Manufacturers have improved silencer technology vastly over the past decade, and cans are getting quieter, more durable, and more affordable. Our selection is easily the best in human history. Still, there are ways to eke out more performance. Here we outline a few ways you can boost your cans' abilities and dispel a few internet rumors along the way.

WHAT'S THAT SOUND?

For the uninformed or misinformed, one of the main reasons to use a suppressor is to mitigate sound and flash. A simple way to think of this is in two parts: first, the rapidly expanding gases that escape the barrel, and then the sonic crack of the bullet traveling down range. Both sounds can be reduced, the gases with the silencer and the crack with subsonic ammunition.

A firearms suppressor works by trapping the expelled gases that follow a bullet leaving the barrel. Chambers, created by baffles or other means, trap the gases,

causing them to cool and dissipate inside the suppressor and preventing or mitigating them as they exit.

It's important to note that "first round pop" (FRP) is also something that occurs when the initial shot burns the existing oxygen inside of a suppressor, causing a noticeably louder first round. Though FRP exists with all normal silencers to some amount, people focus on it more with pistol-caliber and rimfire suppressors.

The Gemtech Aurora II uses wipes to quash the noise. Since wipes are consumables, the Aurora II can go from looking new (above, left) to blown out (above, right) in fairly short order.

WET 'N' WILD

Now that we established silencers take hot gases and cool them, let's talk about ways to allow them to do that more effectively. The most prevalent is making a can "wet" or adding something to the inside of the suppressor to cool the gas quicker. We should note that this practice is *only* meant for rimfire and pistol-caliber suppressors. Rifle cartridges carry much higher pressures and adding liquids to the inside of the suppressor can make for a potentially very unsafe environment — so it's best to leave them out.

Water is an obvious choice for an ablative for a suppressor, though any nonflammable liquid will work (some even swear by Coca-Cola). Placing a tablespoon[ish] amount of water inside the first chamber of the can and swirling it around will help negate the FRP issue, and the next few shots will be quieter. However, the water will burn off quickly, and the suppressor will return to normal function with a magazine. Another popular option is using petroleum jelly inside the suppressor. Once again, about a tablespoon amount in the blast chamber works well. An added benefit is that the jelly will not evaporate or run out of the can, so it stays in place much longer. The petroleum jelly will also take away from FRP problems and last significantly longer than a liquid.

An ablative such as grease, petroleum jelly, or shaving cream (shown here) can significantly reduce first round pop (FRP) with pistol calibers.

As no good deed goes unpunished, the drawback to making a wet can is that the host firearm will get much dirtier, much faster. When petroleum jelly is used, it will seep back into every crevice of the host. Gas blowback and pressure are also significantly increased; this is particularly painful for shooters whose handgun has an optic. Debris will splatter on the optic, and often cause it to be unusable. This can also happen to your eye protection.

With those downsides, why would anyone want to make a can wet, you ask? Well, according to Blake Young, CTO for Primary Weapon Systems (and holder of nine suppressor technology patents), making a suppressor wet can gain anywhere from 1 to 5 decibels of sound reduction on a 9mm can, and even more with rimfire.

WIPING SOUND AWAY

Introducing a substance into the suppressor can help cool the gases faster, but what about keeping the gases in the can more efficiently? This is where the old-school wipe comes into play and, in many cases, really shines. Wipes are a consumable, physical barrier the projectile must first pass through and can tremendously aid in containing the explosive gases.

Wipes aren't a new technology — they've been seen in World War II homebrew suppressors using many types of materials, like leather.

Modern suppressors using this old technique have upped the game with modern materials. The most popular is Neoprene 70A. This is currently used in many models from different manufacturers. The Gemtech Aurora II (full review in RECOIL Issue 37) is a great example of both wipes and petroleum jelly, previously discussed. The wipes are pliable and allow a projectile to pass through and essentially seal behind trapping the gases inside the silencer slightly longer. This concept is similar to self-healing targets. The wipes are scored with an X to help the bullet pass through. Dead Air, Lone Wolf,

An endcap with a disposable wipe (above), such as the Dead Air Ghost-M, can help take the edge off.

The wipe-and-ablative GSL Pill Box is a perfect pair to the Beretta 21A Bobcat chambered in .22LR.

Thompson Machine, Energetic Armament, and numerous others have or currently do use wipes to make suppressors more effective.

Wipes are most commonly used with pistol-caliber and rimfire silencers, but not all. The Energetic Armament Vox S is a rifle suppressor that incorporates a wipe. Pew Science, a newer organization that's revolutionizing suppressor testing, featured the Vox with and without a wipe. Their findings regarding wipe performance with subsonic 300 Blackout are impressive. With the wipe installed, the suppressor received a rating of 58.9dB and without it received a rating of 48.4dB. This may not mean anything to you right now (you should absolutely go check out pewscience.com), but this is a significant increase in performance. Pew Science provides impressive and exhaustive data sets, so if you're any kind of silencer nerd, again, go check them out.

Now for the downsides. Wipes wear out — and quickly. Even with the fabric-reinforced materials, wipes generally last between 20 and 50 rounds. When talking to Mike Pappas of Dead Air Armament, he states, "I think the best application is when subsonic ammo is being used, and it may not be as effective as making a can wet, but it also doesn't create a mess."

As previously mentioned, wipes create a physical barrier the projectile has to pass through. This means that bullets designed to expand may start the process while still inside the can. In the case of the wipe-and-grease-only Gemtech Aurora II, ammunition must be considered. A small silencer like that seems like a great option for a self-defense situation, before you understand only FMJ ammunition can be used. Any type of expanding ammunition will inevitably expand inside the suppressor, causing catastrophic

SPEED OF SOUND

The speed of sound changes with altitude and temperature. The speed of sound is about 761 mph at sea level or approximately 1,100 feet per a second (fps). Bullets traveling slower than 1,100 fps are considered subsonic and will not produce the loud crack down range. Sound is slower at higher altitudes and at lower temperatures.

761 mph
1,100 fps

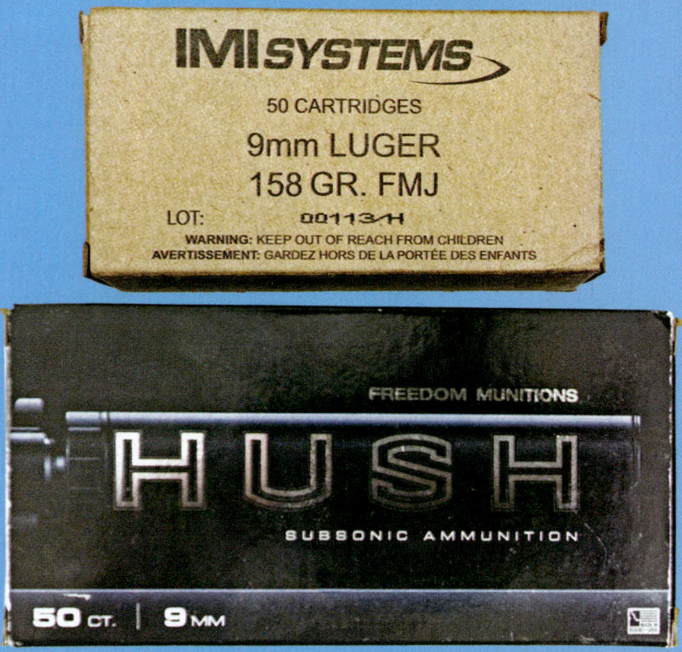

Subsonic ammo makes everything sound great, and heavier subsonics hit with authority. Israeli 158-grain FMJ and 165-grain Freedom Munitions HUSH are some of our go-to rounds if we want to maximize silencer performance.

damage. Accuracy is another topic for wipes. Because the bullet literally touches the wipe, there can be a negative effect on accuracy.

LOOSE ROUNDS

It's possible to assist a suppressor in mitigating sound through outside means. In the old days of Hush Puppies, it's rumored those operatives would purge oxygen from the suppressor with nitrogen and seal the end of the can with a postage stamp to assist in the reduction of FRP. Using dB foam, shaving cream, water, petroleum jelly, etc., can all achieve similar results. Wipes are useful and, when designed into the suppressor, can mitigate some of the downsides; however, also be aware of the safety concerns when using these. **R**

Whether it's petroleum jelly, wipe, water, wire-pull gel, or simply being more choosey with your ammo — you can cut some decibels off.

3 RIFLE BUILDS

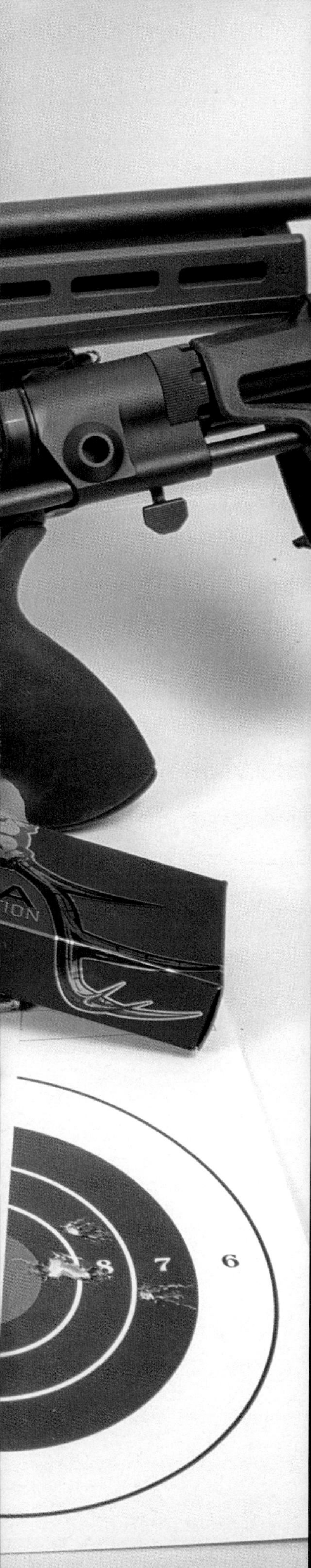

300 BLK BUYER'S GUIDE

Expanding 300 Blackout Subsonic Ammo

By Iain Harrison
Photos by Kenda Lensiegne

300 BLK has a lot going for it, right up to the point at which you try to take advantage of its subsonic prowess for tasks other than punching paper. Common subsonic loads, even if tipped with a fragile match bullet such as the ubiquitous 220-grain Sierra Matchking, plow 0.30-inch diameter holes straight through fleshy targets and keep on truckin' right into the next inconveniently located piece of real estate.

We've tried bullets that in other applications are positively explosive. Hornady's 208-grain A-Max, for example, fragments violently when shoved out the muzzle at even .308 Win velocities, but from a Blackout you might as well be using a DeWalt. Getting a .30-cal to expand reliably at around 1,000 feet per second requires a specialized projectile, and they're not exactly inexpensive.

Being cheap bastards, in times past we experimented by cutting back the jacket tip on an SMK and then boring out the meplat, thinking that this would be enough to induce expansion. Great theory, sucks in practice. Gel testing confirmed that it performs no better than an unaltered bullet, so why waste the effort? Cutting to the chase, we rounded up a small selection of expanding subs. Note there are a few more companies who offer bullets for you to roll your own (Cutting Edge, for example), but as work has displaced gentle-paced pursuits like reloading, we'd often rather pay in money than time.

If you decide that a subsonic 300 BLK fills your requirements for a stealthy hog slayer, or you've settled on that caliber for home defense duties, then you might want to avail yourself of our buyer's guide below.

AMMO	MEAN VELOCITY 16-INCH BBL	ES	MEAN VELOCITY 9-INCH BBL	PENETRATION	EXPANDED DIAMETER	ORIGINAL BULLET WEIGHT	RETAINED WEIGHT	GROUP SIZE
Fort Scott	1,047	39	936	15.5	0.308	190	189.8	2.55
Gorilla	1,035	27	943	17.5	0.933	205	204.9	0.88
Hornady	1,150	26	1,017	18	0.585	190	170.2	0.75
Lehigh	1,150	49	1,034	21	0.550	194	180.5	0.74
Noveske	1,177	22	1,004	17	0.551	220	190.8	1.35
SSA	1,156	29	1,022	14	0.468	220	187.8	1.32

FORT SCOTT 190-GRAIN SCS BRUSH HOG $1.30 / ROUND

Ever heard of these guys? Neither had we, until we walked into a gun store near San Antonio, Texas, where the counter jockeys assured us they were well regarded by local hog hunters. Intrigued, we bought a couple of boxes for testing. Unlike the rest of the projectiles in this test, these are non-expanding, relying for their wounding abilities on becoming unstable in media more dense than air. Gasp! The fabled M16 tumbling bullet!

Despite a healthy dose of skepticism, they turned in a decent performance, coming to rest after 180 degrees of flippage and 15.5 inches of penetration in bare gel, with the bullet's base just barely under the back end of the block. Fort Scott seems to be using a set of dies that diverges from industry norms, which resulted in a misfire rate of around 50 percent in our bolt gun — the only sample to do so. On examining the case shoulder (which, in 300 BLK, is even less significant than Chuck Schumer's dick) it would appear that it's both steeper and shorter than competitive brands, allowing the case just enough room to shift forward under firing pin impact. It ran fine in our AR, though.

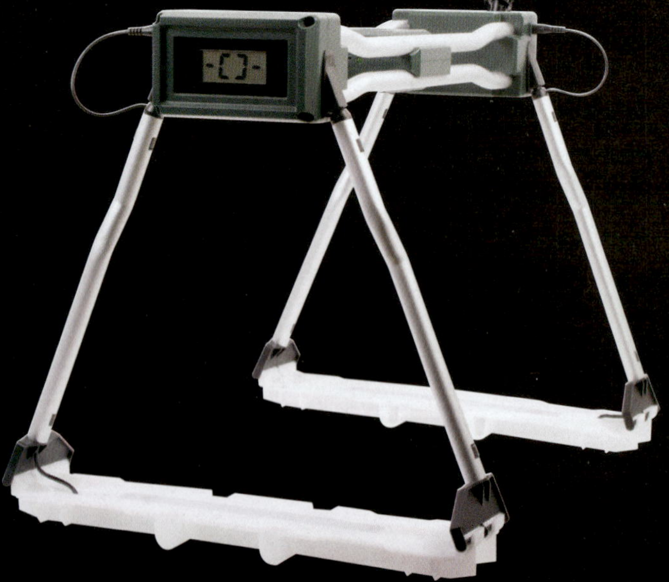

TESTING PROTOCOL

All loads were tested for velocity using both 16- and 9-inch barreled weapons. A Remington 700, equipped with a Nightforce ATACR 4-16x42 and SilencerCo Omega was employed as our longer rifle, and it was this combo that we also used for accuracy and gel testing. Five-shot groups were shot at 50, rather than 100, yards as we felt this was a realistic distance for subsonic use. The shorty gun was our budget hog hammer — a franken AR built on a Gibbz side charging upper and equipped with a Maxim Defense PDW pistol brace. Note that all loads except the Fort Scott and Gorilla offerings were supersonic from the 16-inch barrel.

Velocity readings were taken using a Caldwell G2 chronograph, which has the useful ability to store data on your phone via Bluetooth. Chrono was set up 10 feet from the muzzle, ambient temp was 69 degrees at 2,100 feet elevation, and gel testing was performed using two, 16-inch Clear Ballistics gel blocks, placed end to end.

GORILLA AMMO 205-GRAIN SILVERBACK $1.95 / ROUND

Gorilla's monolithic projectile features a gaping hollow point, which initiates pretty healthy expansion after about 3 inches of penetration in gel. One of the two loads achieving subsonic velocities from our 16-inch bolt gun, it's mouse-fart quiet even in longer barrels. Not only did it retain all its weight after impact, but petals on the copper bullet remained sharp enough to nick careless fingers when pulling it from the gel block. Accuracy was plenty good for dispatching critters at subsonic ranges.

HORNADY 190-GRAIN SUB-X $1.15 / ROUND

Announced in December 2017 and released for sale three months later, Hornady added an expanding subsonic bullet to its existing five Blackout offerings.

Built to their usual standards, the bullet features a red polymer tip to initiate expansion, a detail no doubt familiar to users of their Critical Duty line of pistol ammo. In our test rifle, it proved to be an accurate and consistent performer, and if a lone flyer were discounted, would've turned in a 0.36-inch group. Of the conventional jacket-and-core projectiles in this test, it was the only one not to separate. Note, however, that some polymer mags may have problems in presenting the blunt tip high enough to make it up the feed ramp in your favorite semi auto. Either use regular old GI versions or Magpul's 300 BLK-specific offering.

LEHIGH DEFENSE 194-GRAIN MAXIMUM EXPANSION $2 / ROUND

Lehigh's wicked-looking solid copper bullets turned in an admirable performance, expanding as advertised and growing sharp, symmetrical talons when shot into gel. When dug out, they showed evidence that the fast twist barrels of our test guns would've caused additional tissue damage, as they screwed their way through the medium — in this case making three complete revolutions before coming to rest after shedding a petal. Those wanting additional penetration should consider Lehigh's loading; it exceeds the FBI protocol, but if you encounter bigger hogs it affords a margin of safety on quartering shots.

NOVESKE / NOSLER 220-GRAIN BALLISTIC TIP $2.25 / ROUND

Averaging 20-fps faster than the otherwise similar SSA load, this collaboration between two of Oregon's most respected gun companies is unmistakable with its ominously dark finish. Accuracy was good enough for the task at hand, and the bullet core achieved 17 inches of penetration, after parting company with the remaining jacket around the 15-inch mark. About 30-grains worth of ballistic payload was shed in total, leaving a smoky-looking trail through our gel block as it threw off particles of lead and copper.

SSA 220-GRAIN BALLISTIC TIP $1.80 / ROUND

Silver State Ammo was acquired by Nosler in 2013, and we suspect that both this load and the Noveske/Nosler above use the same bullet, with a different polymer tip and coating. Group size was almost identical to the Nosler load, and both bullets separated from their remaining jackets after 12 to 14 inches of penetration in bare gel. As this meets the FBI requirements, we don't see this as a negative for defensive use, because both resulting fragments penetrate sufficiently to reach vital organs. R

JUST BECAUSE YOU CAN

The Gear Head Works One 300BLK Bolt Action Pistol

By Steven Kuo

Some of the coolest products come about because someone was tinkering in the workshop, making something purely for their own enjoyment. Such is the case with Gear Head Works' creation, a stubby little bolt-action pistol called the One.

Do you remember the Remington XP100, first introduced in the '60s? Based on a 40X action, it was a unique bolt-action pistol that found ardent fans amongst handgun hunters and target shooters. Most variants were single shot, but the XP100-R version had a four-round internal magazine. Remington even developed the 221 Fireball cartridge for it, and it certainly earned its nickname. Like the stock-less AR-15 pistols of old, the XP100 was hardly the easiest gun to shoot effectively.

Gear Head Works is best known for the compact and very rigid Tailhook pistol brace, which like all modern pistol braces have been a revelation when employed on AR-15 pistols and the like, making the clumsy weapons eminently practical without the need for a tax stamp.

Paul Reavis, the founder of Gear Head Works, has a history of dreaming up and constructing unique builds purely for his own amusement. Four years ago, he came across some virgin Remington Model 7 actions, which had yet to be built into rifles. Thus, it occurred to him that he could do a pistol-brace-build with a bolt-action platform. If you have a virgin action, you can elect to make a pistol with it from the get-go, attach a brace, and avoid ending up with an SBR.

Reavis already had some 300BLK bolt-action rifles that he enjoyed shooting and hunting with, but he wanted to make a pistol version that he could fit in a regular-sized backpack and would be a great suppressor host. He started by installing a 12-inch barrel and an MDT LSS chassis, affixing a Tailhook pistol brace with a hinge to fold it on the left side.

GEAR HEAD WORKS

But he wanted to make it shorter and lighter, so he put it on a keto diet, courtesy of a Bridgeport mill. He chopped it down and lightened it as much as he could, then cut the barrel to leave the threads protruding just past the handguard. He also swapped the hinge to fold on the right side and cut a hole in the stock tube for the bolt handle. The result was like that family friend whom your parents wanted you to marry — functional, but not particularly attractive. So to attach the Tailhook to the hinge, Reavis machined a custom tube that tapers between the mismatched outer diameters and is hollowed out throughout its entire length to clear the bolt handle when folded. After a bit more tweaking, he put a suppressor on it and had a cool new toy for the range; he took it hog hunting as well.

Reavis enjoyed shooting it, especially suppressed with subsonic ammo. But he hadn't intended on producing and selling the little pistol. SHOT Show 2018 was approaching, and he wanted another gun to demonstrate the Tailhook pistol brace at his booth that wasn't a typical AR-15. So he took it to the show. People were so enamored with the little bolt gun at SHOT and at the NRA show in the summer that Gear Head Works decided to put it into production, naming it the One.

THE ONE

We tested the original prototype gun, which was built entirely by hand. The production models will differ in several ways from the gun you see here:

› The production gun will be built on Remington Model 700 actions rather than Model 7 actions. A fluted bolt will be optional.

› The custom MDT LSS chassis will have M-LOK slots on the sides and bottom, quick detach sling swivel sockets, and additional lightening. It'll weigh less than a pound stripped, about 6 to 8 ounces less than a standard LSS chassis. The MDT chassis takes AICS-pattern .223 magazines.

› The gun will come standard with a non-folding Mod 2 Tailhook brace that telescopes. A fixed Mod 1 Tailhook and folding brace tube (shown here) will be optional and fold to the left rather than right side, by popular demand from customers.

› The barrel, chambered in 300BLK, will be 9 inches long versus 8.4 inches on the prototype, with spiral fluting available as an upgrade. Both are threaded ⅝-24.

› The 20MOA scope base will be optional.

› The carbon fiber Venom Defense grip will also be optional; the gun will come with a Magpul K grip standard.

› The gun will come with a solid Cerakote finish, with various camouflage patterns available as an upgrade. Shown here is ULTerra Camo's "Fragment" pattern, done by BAM Custom in Tennessee.

The gun was outfitted as closely

Folded, the One is incredibly compact. It fits perfectly and discreetly in the new 5.11 Tactical AMP24 backpack in a gray-man gray color.

as possible to the original intent. The low-power variable Leupold 1.5-4x scope is compact, mounts low for a better cheek weld, and features an illuminated Firedot reticle and a good zoom range for expected engagement distances. It sits on a 20 MOA rail to provide more elevation adjustment for the mortar-like 300 BLK subsonic trajectories. The Atlas bipod is compact, but rock solid. With the Tailhook extended, the total length was 27 inches with a thread protector.

To quiet the beast, we obtained an early sample of Dead Air Armament's brand-new Nomad-30, a perfect match for the One. Lightweight, quiet, and versatile, it's an excellent balance between performance, size, weight, and price. We installed a Dead Air Keymount muzzle brake on the gun to provide quick-detach convenience for transport.

Folded, the Gear Head Works pistol is just 18.1 inches long with a thread protector and 20.3 inches long with the Dead Air Keymount. It fit perfectly in the new 5.11 Tactical AMP24 backpack, which also has stretchy internal side pockets sized for water bottles that are a perfect spot to stash the Nomad. All that hardware adds almost 9 pounds of weight to the pack, but in a discreet tungsten color, the whole package virtually disappears on your back or in your trunk.

AT THE RANGE

The One is a pretty compact weapon, so there isn't much real estate in front of the magazine well and the length of pull with the non-telescoping Mod 1 stock is 13.25 inches. Still, smaller guys found the ergonomics to be comfortable. Bigger guys felt cramped, but that's the case with pretty much any pistol brace-equipped gun.

Fit and finish were excellent, and you wouldn't know that the prototype was built with a manual mill. The chassis, hinge, and Tailhook brace blended together seamlessly.

The Mod 1 Tailhook folds neatly on the right side on the prototype gun. By popular demand, the production gun will fold on the left side.

We liked having the folder close on the right side of the gun — we weren't bothered by not being able to cycle the action with the brace folded and appreciated a tidier package when folded, with the tube laying over the bolt handle. However, Gear Head Works is deferring to customer feedback, which was much more in favor of folding on the left side. The hinge on the prototype didn't lock closed and, like your typical congressional representative, swung back and forth freely, which was an annoyance. While the production gun won't lock closed either, we're told that the hinge will be much tighter so that the brace won't flop open on its own.

Getting behind the gun, it definitely helps to use a low-mounted optic for a better cheek weld. As you might expect, in prone, even small guys felt cramped behind the glass. In other supported and unsupported positions, it wasn't too hard to find a reasonably comfortable shooting position. The Tailhook brace was extremely rigid, though its skeletonized L-shape was certainly not as useful as full-featured rifle stocks. The Atlas bipod proved very stable, as always, and Gear Head Works added a nice contour in the chassis behind the bolt handle that's perfect for the thumb of your shooting hand, for those who place their thumb on the strong side of the weapon.

The Remington trigger was clean, but heavy for our taste. The external adjustment screw is easily accessed via the skeletonized trigger guard, but we were only able to adjust it from about 5 to 4 pounds. We're trigger snobs, so we'd plan to swap in an aftermarket trigger, such as a Geissele (see Incoming on page 22) or Timney.

We tested 200-grain subsonic and 125-grain supersonic 300BLK loads from Maker Bullets. Both feature CNC-machined solid copper bullets, designed to expand effectively and to be used with suppres-

You may feel cramped behind the gun, though you'll feel that way with any pistol-brace-equipped gun. Otherwise, ergonomics are good, with a nice contour on the chassis behind the bolt handle for the thumb of your shooting hand.

sors. Out of the 8.4-inch barrel, the subsonic loads posted an average muzzle velocity of 993 fps suppressed, with a standard deviation of 11. The supers averaged 1,830 fps with a standard deviation of 18. Unsuppressed, muzzle velocities dropped just 5 to 10 fps. Muzzle velocities were measured with our trusty Magnetospeed.

We shot groups from a bench, achieving 1.5 to 2 MOA five-shot groups with supers and subs. Zero shift with and without the Dead Air Nomad-30 was around ½ MOA. Reavis told us that the prototype, originally for his own personal use, was built with the cheapest barrel he had on hand at the time. The production gun will utilize high-quality chromoly barrels with button rifling and 1:7 twist. They'll also feature a lighter profile and optional spiral fluting. He expects them to provide better precision, which we like to see in a custom bolt gun. Still, even a 2-MOA 300 BLK gun will put down hogs and steel just fine, and it can take some experimentation to identify loads that work particularly well in your gun. Not to mention that a 200-grain pill at 1,000 fps with a 100-yard zero will drop over 100 inches at 300 yards.

Speaking of subsonic, the One really comes into its own shooting subs with a silencer. Cranking the Nomad-30 onto the muzzle made it feel like shooting a pellet gun. The combination of the bolt-action platform, subsonic ammo, and the effectiveness of the Nomad-30 resulted in minimal felt recoil and a very subdued report. The One was an absolute blast on the range, punching paper and ringing steel. It's a real attention-getter too, turning a lot of heads at the range during our testing. It'd be a great truck gun, and we can't wait to take it on a hunt.

Some cynical folks say that "just because you can, doesn't mean you should." In this case, we're really happy that they did. If you're offended by the idea of the One, as some commenters on social media seem to be, we'd suggest that you send a few rounds downrange with it before you issue a final verdict — it would even get a rise out of Lord Varys. It's expected to be available by the end of the year, with a projected retail price of $1,499 for the base gun. We'll take One, with a can. **R**

Super and subsonic ammo from Maker Bullets feature CNC-machined solid copper bullets. Zero shift with and without the Dead Air Nomad-30 was around 1/2 MOA. With the can installed, the subs were quiet and soft-shooting.

GEAR HEAD WORKS ONE

Caliber: 300 BLK

Barrel Length: 9 inches (8.4 inches, as tested)

Overall Length: 27.6 inches extended, 18.7 inches folded (0.6-inch shorter, as tested)

Weight (Unloaded): 5.7 pounds (6.0 pounds as tested)

Magazine Capacity: 10 (AICS-pattern .223 magazines)

MSRP: $1,499

URL: www.gearheadworks.com

Accessories:
Folding Tailhook Mod 1 option ($120)
Camouflage finish option ($150)
Venom carbon fiber grip option ($20)
20MOA scope mount (TBD)
Leupold Mark AR Mod 1 1.5-4x20mm Firedot scope ($585)
Weaver scope rings ($15)
Atlas bipod with ADM quick detach base ($280)
Dead Air Nomad-30 silencer ($916)
Dead Air Keymount muzzle brake ($89)
5.11 Tactical AMP24 backpack ($190)
5.11 Tactical Admin gear set ($35)

Price as featured: $3,899

THE RIFLE THAT wish BUILT

We Build a Gun With Parts From the Worst Vendor on the Internet

By Dave Merrill

Over the years, we've seen all sorts of crazy cheap deals on parts and pieces; some of them with questionable origins. The perfect place to put them all together was right here in our budget issue. Who needs Black Friday when you can build yourself a working rifle from the cheesiest of Chinesium from Wish.com — or at least try to.

We did our best to avoid products that were simply stolen intellectual property from American companies. Despite our best efforts, the iron sights we purchased ended up being garbage knockoffs of Magpul MBUS sights — what was pictured in the listing was totally different. Still, they ended up being absolutely worthless due to two reasons: The apertures freely rotated, and the handguard was total trash.

PARTS AND PIECES

When the handguard arrived, it actually looked pretty decent. Not only was it lacking that chalkboard-made-in-China feel, it had anti-rotational tabs, a barrel nut made of steel, and a whopping five quick-detach sling points. If you put it on a shelf in a gun shop, no one would believe it only cost $27. Ultimately, it ended up being LG;FU. That is to say: Looks Good; F*cked Up.

None of the QD points were in-spec; all were undersized. When the barrel nut was properly aligned to allow the gas tube to enter the upper receiver, the rail was canted several degrees. We had to shave down the alignment tabs to line up the rail with the upper. The handguard itself was a study in poor aluminum extrusion; rather than remaining in-line with the upper from tip to tail, it nearly touched the barrel at the

muzzle end. Even if the Magpul-knockoff iron sights were OK, the rail itself precluded their use.

One of the few components we couldn't procure from Wish was the barrel. No matter, there are several über-budget options. We went with a Bear Creek Armory blemished barrel. We know what you're thinking, but no, not all of them are blems, apparently. This mid-length 5.56mm barrel had a gas port mic'd at 0.088 inch — too large to start out with.

The lower receiver was an Anderson Arms purchased on sale for $25, and the upper receiver a blem from Gorilla Machining with the original intended manufacturer's markings milled off of the side. We skipped the ejection port cover, because we didn't want to spend the extra $2.

The gas block, shockingly, was made from steel. Instead of a press-on, it's the far inferior clamp-on type that went out of favor more than a decade ago because of gas leaks. The finish came off when you looked at it.

Though sold on Wish, the listing for the bolt carrier group claimed it was U.S.-made from a company called KM Tactical. A cursory web search brought up a company based in Missouri with an online store selling a lot of similar items. As expected, it wasn't marked as high-pressure-tested — not that we'd believe it if it were.

The lower parts kit was a mere $20, and the included springs looked like it. Surprisingly, it also included a winter trigger guard and an ambidextrous safety selector. The only part we were missing for this build was a pistol grip, so we pulled a black Hogue from a parts box.

OUTFITTING

Since the iron sights were out, we decided not to buy a complete fake on Wish and turned to Vortex for an inexpensive red-dot sight. We chose a SPARC AR that runs on a single AAA battery. This would ultimately be the highest quality item that would ever touch this rifle. While we believe a weapon mounted light to be paramount for any gun configured for defensive purposes, we skipped it here because we couldn't imagine a single scenario where we'd stake our lives on this rifle.

It's not lost on us that the MSRP of the Vortex SPARC AR exceeds the rest of the rifle combined.

AT THE RANGE

The rifle experienced a 50-percent failure rate right from the start — frankly, better than we expected. The main culprit was that the trigger wasn't resetting. Under normal circumstances you'd check for incorrect hammer spring installation, but in this case, it was because the springs were made of zinc or butter.

Once that was corrected, the rifle actually ran. As to how long it'll continue to run? Experience tells us it won't hold a candle to a properly built, in-spec rifle. But we'll continue to flog it and chronicle the misadventures on RECOILweb. Because we're masochists.

LOOSE ROUNDS

The fact is, a cheap build may not end up being as cheap as you thought. Firstly, none of the prices reflected in this article showcase how much shipping actually costs, nor how much you'll wait for your wares. Additionally, a handful of the subpar items we received had to be replaced entirely, adding to the overall time and cost. Furthermore, we received parts that didn't match the descriptions, and we have a handful of well-known failure points in this rifle as well.

We did give this rifle the worst paint job we can think of. So, there's that.

Ultimately, it was an interesting thought experiment, but what we received wasn't all gravy. You'd be better off looking for a sale on an inexpensive AR, such as a Palmetto State Armory, rather than trying to piece everything together from a sketchy foreign vendor. Unless you just want to make your partner cry and your gunsmith rich. **R**

COMPONENTS	MSRP
Anderson Manufacturing lower receiver	$25
Gorilla Machining blemished upper	$25
Bear Creek Arsenal blemished barrel	$45
Complete buttstock assembly	$24
Gas block and gas tube	$14
KM Tactical NiB bolt carrier group	$68
Handguard	$27
Charging handle	$8
Lower parts kit	$20
TOTAL:	$257
ADDITIONAL COMPONENTS	
Vortex SPARC AR	$259
Black Aces 5.56 Po' Boy Silencer	$199
PRICE AS CONFIGURED:	$715

"YOU'D BE BETTER OFF LOOKING FOR A SALE ON AN INEXPENSIVE AR, SUCH AS A PALMETTO STATE ARMORY, RATHER THAN TRYING TO PIECE EVERYTHING TOGETHER FROM A SKETCHY FOREIGN VENDOR."

GOING OFFGRID

A Multipurpose Survivalist SBR Themed Around Our Sister Publication

By Patrick McCarthy

There seems to be a nebulous atmosphere of gloom and doom circulating in America today. Politicians and TV pundits claim our government is on the brink of collapse; *Time* magazine's scowling teen slacktivist-of-the-year thinks we'll all be fighting over newly formed beachfront property soon; keyboard commandos on social media are certain of a forthcoming civil war. Any rational person takes these pessimistic perspectives with a huge grain of salt. Still, it's never a bad idea to insulate yourself from a variety of worst-case scenarios according to the essential mantra of preparedness gear, *better to have it and not need it than to need it and not have it.*

With this in mind, we set out to build a rifle that might serve as a one-size-fits-most tool for the types of situations discussed in our sister publication, RECOIL OFFGRID. It might need to fit the role of bedside home-defense implement, bug-out backpack gun, and/or a means of putting food on the table. We therefore decided to go with an SBR for maneuverability, a folding stock adapter for packability, and a silencer to save our hearing in settings where active ear pro might be a luxury we don't have. Reliability with a variety of ammunition, with or without the silencer, was also prioritized to enable scrounging and scavenging.

The build began with a matched lightweight billet receiver set from Ascend Armory. It features numerous cuts and recesses to shave excess ounces, and comes with preinstalled ambidextrous bolt catch, billet takedown pins, and unidirectional threaded trigger pins. Both the upper and lower were laser-etched with the RECOIL OFFGRID logo. Ascend also provided a billet safety selector, but an apparent tolerance stacking issue caused it to lock up with the drop-in single-stage Velocity trigger; a Battle Arms Development selector nicely circumvented this issue.

Next, we sourced a Quickmount flash hider and titanium Shield silencer from Gemtech. This setup dramatically tames the report of the weapon without causing it to feel unwieldy.

We selected an 11.5-inch carbine barrel from Sionics with the standard gas port size — unlike the company's reduced gas port options, this enables the rifle to cycle just as reliably without the can. The crew at Sionics assisted with assembly, includ-

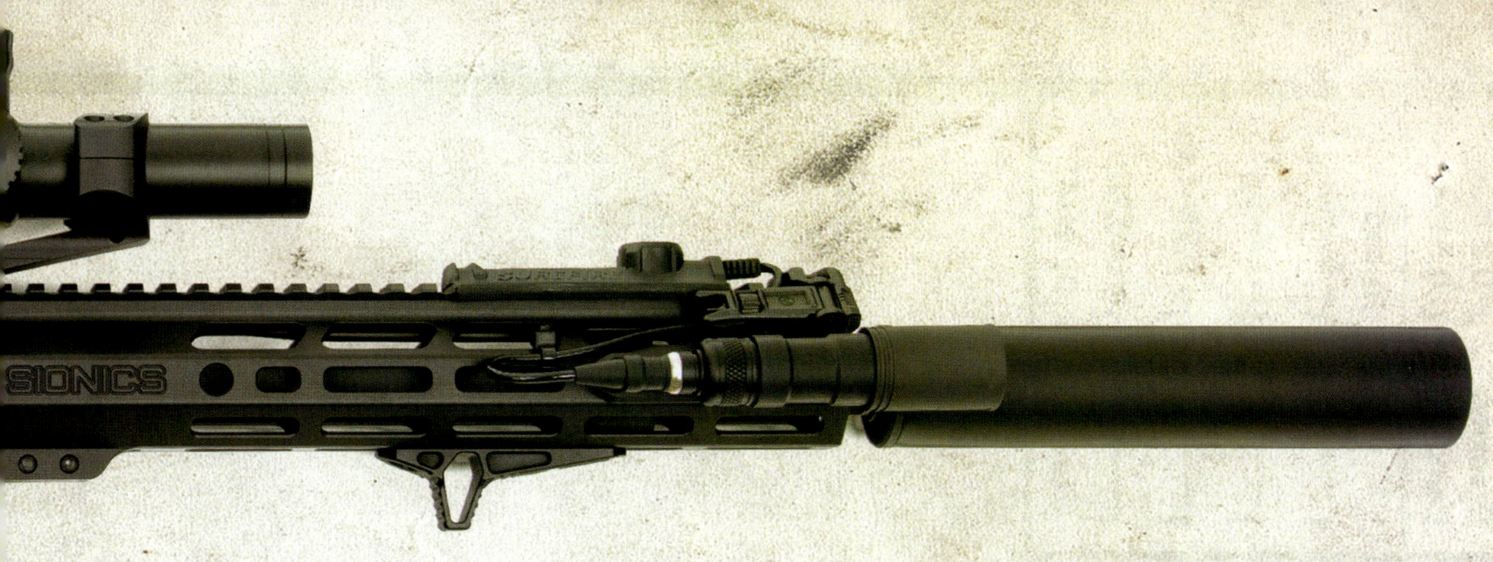

ing installation of the M-LOK rail, which required light lapping of the indexing tabs with a diamond file to fit the billet upper. The upper was finished with a Sionics BCG with easy-to-clean NP3 coating and a VLTOR charging handle.

In order to make this SBR even more transportable, we installed a Law Tactical folding adapter. With the silencer removed and stock folded, this brings the gun's packable length to less than 23 inches — just right for slipping it into a backpack or under a truck seat at the cost of an additional 10.5 ounces.

The VLTOR A5 buffer system was chosen as a means of fine-tuning the action, but it also has the pleasant side effect of smoothing out the recoil impulse. With the standard spring and A5H2 buffer, we experienced some occasional short-stroking. Before playing with buffer weights, we dropped in a SOLGW green spring, which eliminated the issue.

A Magpul MOE-SL stock keeps the rifle slim while it's folded, and an MOE+ pistol grip houses a container of CLP for field maintenance. Up front, we installed an Ascend Armory billet AFG for added control. No home-defense-ready carbine is complete without an illumination device — for this build, we sourced an Arisaka Defense 300 Series light that produces a 325-lumen, 23,000-candela spot beam. Paired with an Arisaka inline mount, this light tucks in close to the muzzle. It's also compatible with SureFire accessories, such as the Scout tail cap and SR tape switch seen here.

Finally, we knew the theme of this build required a versatile optic as well as an easily accessible backup iron sight setup. This Nikon 1-4x LPVO fulfills the first requirement, with an illuminated reticle that's effective for close-range engagement. The second need was met by way of Magpul MBUS Pro Offset sights, which can be folded away to sit flush with the upper when they're not needed.

Despite what clickbait blog articles may claim, there's no such thing as an all-purpose SHTF gun. Even if you're a distant relative of Nostradamus, you can't predict the exact challenges you might face in the future, and even if you could, there's no single weapon configuration that's optimal for all of them. What you can do is make educated guesses about the most likely scenarios that apply to your location and lifestyle, build accordingly, and train frequently to match your hardware with the essential software. Nothing dispels gloom and doom faster than confident preparedness. R

PARTS	MSRP
Ascend Armory Lightweight Billet Receiver Set	$610
Sionics 11.5" Chrome-Lined Carbine Barrel with Low-Profile Gas Block	$220
Gemtech 5.56 Flash Hider	$100
Gemtech Shield Silencer	$995
Sionics 10.5" M-LOK V3 Handguard	$150
Ascend Armory Angled Foregrip	$39
Arisaka Defense 300 Series Light	$175
Arisaka Defense Inline Scout Mount M-LOK	$40
SureFire UE Tail Cap and SR07 Switch	$174
Sionics NP3-Coated Bolt Carrier Group	$190
VLTOR Mod 4 Charging Handle	$53
Battle Arms Development BAD Ambi Safety Selector	$60
Velocity Triggers MPC Straight with Finger Stop, Nickel Coated, 3 lb.	$190
VLTOR A5 Receiver Extension and A5H2 Buffer	$95
Sons of Liberty Gun Works A5 Green Spring	$20
Law Tactical Folding Stock Adapter Gen 3-M	$270
Ascend Armory QD End Plate and Titanium Enhanced Castle Nut	$40
Magpul MOE-SL Stock	$60
Magpul MOE+ Grip with Lube Bottle Core	$39
Magpul MBUS Pro Offset Iron Sights	$190
Magpul MS1 Padded Sling with QDM Swivels	$116
Nikon Black Force 1000 1-4x24mm	$400
Warne MSR Mount	$137
Total	$4,363

We Make Short Things Shorter

By Dave Merrill

HACKSAW SPECIAL

The build for this issue is all about small. And if something wasn't small enough, we followed the best traditions of the Wile E. Coyote School of Gunsmithing and busted out the hacksaw and Dremel.

The center of this story revolves around an 8.5-inch Rainier Arms Ultramatch Mod2 barrel chambered in .223 Wylde. Most everything about this barrel is built to Rainier specs: Starting with a Melonited 416 stainless barrel with a 1:7.5 twist barrel, a TiN barrel extension is added, and the barrel is atypically contoured and profiled. Instead of a standard 0.750, skinny 0.625, fatty 0.875, or 0.936-inch gas block seat, Rainier Arms determined 0.800 inch was ideal for their purposes. Therefore, a Rainier Arms gas block must be used, which they'll happily bundle for you. Unfortunately, we didn't end up with an adjustable block so we needed to ensure we used a lower-pressure silencer.

While a stubby barrel might seem contrary to the concept of "Ultramatch," mostly what you lose with a short barrel is velocity and not accuracy. You do need to be more choosey with ammunition, though.

The trigger follows the same match theme. The Hiperfire X2S Mod-1 is a two-stage trigger with a slightly longer first stage for those who like to pull up some slack before a crisp 3-pound break. If you've never installed a Hiperfire before, we suggest you watch one of their videos on the subject because it's significantly different than a standard trigger.

To keep everything short and tight, we dropped on the spring-loaded Strike Industries PDW stock, and it's also fun to deploy.

We had a 15-inch Bootleg Inc CamLock M-LOK handguard, which we initially thought we would dock a silencer inside since the rail is quick-disconnect. However, with an internal diameter of under

1.4 inches, virtually no 5.56 suppressor would fit. Plus, it makes everything hot and horrible. A hacksaw gave us a custom size, and a Dremel and Alumablack cleaned up some of our rough work.

The upper receiver is also by Bootleg Inc, and we popped in a complete Ballistic Advantage bolt carrier group. The lower receiver was robbed from an already-registered SBR because the ATF was taking their sweet time as this was being put together. The charging handle is the Battle Arms Development RACK, which is a totally different approach to an ambidextrous handle. Left side, right side, weird angles — it doesn't matter. The RACK is a smooth operator.

Most all of the small parts and pieces also came from Battle Arms Development. We combined their enhanced magazine release with their enhanced modular magazine button.

For the selector, we rolled with the BAD-ASS PRO ambidextrous safety selector. It doesn't have any pins or screws to lose, and it can be configured for either a 60- or 90-degree throw. This was the second part that got the chop to for reasons very specific to me. Since I use a very high hold on the pistol grip and have what have been described as child's hands, even short throw ambi safeties can dig into my trigger finger when moving to the firing position. A little shortening and now all is well.

The G10 grip is a VZ Operator II Gen 2. Out of the box, these are extremely aggressive, so a little sandpaper softened it a scootch, and once again a hacksaw made it into a stubby Kurz grip.

For a silencer, we initially thought a Dead Air Nomad-Ti (see page 150) would make for a lightweight warrior, even though a shorty 5.56 barrel would eat it in short order. But unlike other Dead Air cans, there are barrel restrictions with the Nomad-Ti (no shorter than 12.5 inches with 5.56). We decided instead on a SilencerCo Chimera, which has no such restrictions. We added a Dead Air Saker KeyMo adapter and flash hider to round out the package.

We pondered the optic configuration for a considerable period. While purportedly this could fill a more precise role in a pinch, we decided on a magnifier setup instead of a low-powered variable optic for this little guy for best use in close quarters. A tan EOtech EXPS3-0 was installed, and the latest flipside 3x EO magnifier, the G43, was placed behind it.

For a weapon light setup, we scoured the box of lights in the shop and chose an older Streamlight ProTac Rail Mount 2 on an Arisaka Defense in-line mount with a Cloud Defensive LCSmk2k mount for easy use on the 12 o'clock rail.

Because this is a hacked-down handguard complete with a pistol-length system, the forend can get super hot. To negate this, we enlisted the help of Arcane Concerted and their T.A.R. grip rail covers. Available in several colors and lengths, these protect your hands from heat while leaving the top rail open and can be customized for cutouts such as VFGs or other accessories.

Recoil is robust for a 5.56, so we'll be addressing that in the near future with some form of gas regulation. In the end, we ended up with a relatively lightweight, certainly small, fast, and accurate rifle. And that's something we can all get behind. R

PARTS	MSRP
8.5-inch Rainier Arms Ultramatch Mod2 barrel	$270
Bootleg Inc upper receiver	$180
Noreen Firearms billet lower	$100
Bootleg Inc CamLock Handguard	$160
Strike Industries PDW stock	$275
Hiperfire X2S Mod-1 trigger	$200
Ballistic Advantage BCG	$125
BAD-ASS-PRO selector	$45
BAD enhanced mag catch	$9
BAD-EMMR mag release	$24
Battle Arms enhanced bolt catch	$15
Battle Arms RACK charging handle	$125
VZ Operator II grip	$85
SilencerCo Chimera	$1,030
Dead Air Saker KeyMo	$249
Dead Air flash hider	$89
Arcane Concerted T.A.R. grip	$90
Streamlight ProTac RM2	$132
Arisaka Defense inline Scout mount	$40
Cloud Defensive LCSmk2k	$70
EOTech EXPS3-0	$725
EOTech G43 magnifier	$629
My Southern Tactical Crayon PMag	$28
Total	**$4,695**

The ATAC Defense ADER with the Dead Air Nomad-Ti

BY DAVE MERRILL

TUNGSTEN & TITANIUM

You may not have ever heard of ATAC Defense before, but you've definitely seen their wares. This Mississippi-based company started out more than two decades ago with automotive parts manufacturing before moving into the firearms realm eight years ago. Automotive and aviation manufacturing companies transitioning to firearms has become something of a common refrain over the years; it's almost like a natural progression. For nearly a decade, ATAC Defense was the man behind the curtain, OEMing more than 300 parts for different companies, but now they've decided to make their public debut under their own name.

THE ADER

ATAC has three different initial offerings, all with their own variations. Featured here is the ADER: the ATAC Defense Enhanced Rifle. It has all of the bells and whistles, while the ADBR (ATAC Defense Basic Rifle) is for those just getting their feet wet. They also have 9mm pistol builds with 4.5- and 8.5-inch barrels. All are available with a wide variety of Cerakote finishes, and the enhanced models have several triggers to choose from.

The ADER is crammed with custom parts. Many companies call their rifles ambidextrous when there's actually nothing ambidextrous going on other than the safety selector. Not so with the ADER, which has an enhanced and ambidextrous mag release, ambi safety, and oversized and ambi charging handle — all of ATAC's own design. Even though the safety has a 90-degree throw instead of the increasingly common 45- or 60-degree short throw, it won't rub against the trigger finger when using a high hold (a common issue for me personally).

ATAC offers three different triggers: a 3.5-pound single-stage with either a straight or curved bow, or a two-stage with a 1.5-pound first stage and 2-pound break. The two-stage ATAC trigger is a total standout — among the very best produced for an AR-15. If ATAC Defense ever decides to sell these separately, you should snatch one up.

If ATAC Defense ever decides to sell their controls or triggers separately — you'd be well-advised to snatch them up.

Other specifications get high marks for quality too. HPT/MPI bolts cut from Carpenter 158 steel. 1/7 twist barrels formed from 4150 steel blanks per Mil-B-1159SE specifications and then black nitrided. The entire bolt carrier group is also NiB treated.

The receivers themselves are basic 7075 forged affairs. In the future, we'd like to see integral trigger guards and an ambidextrous bolt lock/release à la ADM, Grey Ghost Precision, and Knight's Armament.

For furniture, ATAC turned to Mission First Tactical, the other polymer accessory company, for their Minimalist stock and Engage grip. ATAC's own M-LOK handguard has a continuous top rail and integral QD sockets.

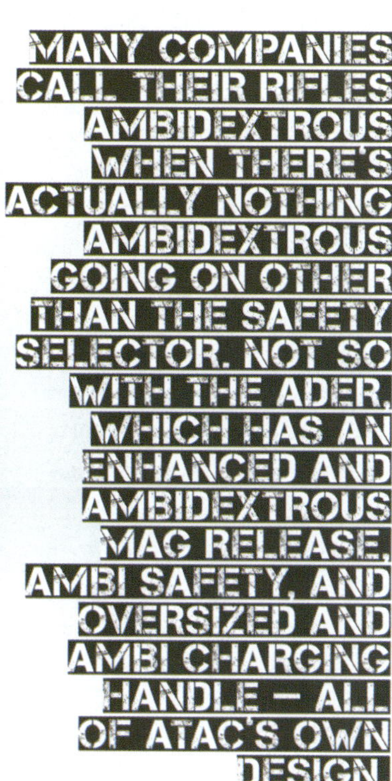

The gas system may give you pause, as the ATAC Defense ADER features a 16-inch barrel with a carbine-length gas system. However, it's not an immediate deal-breaker; while mid-length gas systems generally make for a soft shooter, carbine gas systems are incredibly reliable. ATAC says that longer gas systems will be an option in the future. We'd like to see ATAC implement an option for an adjustable gas block as well.

The ADER ships with a three-chamber brake that curiously features external threads at the base. We asked ATAC about plans for their own suppressor line in the future; it was a little hush-hush but with a wink and a nod. While the brake is effective, it's also very loud just like all other brakes. We removed it to affix a Dead Air Nomad.

THE NOMAD-TI

Dead Air Silencers first released the Nomad in October of 2018. This was Dead Air's first wide-body, tubeless, 1.735-inch diameter can with the now-universal HUB 1.375x24 TPI rear end. Therefore, a nearly endless supply of mounting solutions is available.

The Nomad isn't just a lower-pressure suppressor due to sheer volume — just look at that X-ray. It's clear the Nomad isn't simply the same old silencer design in a slightly fatter profile. Each baffle has its own secondary chamber to store and then bleed gas after the projectile passes, effectively making the Nomad a silencer with multiple bore evacuators. We first covered bore evacuators in RECOIL Issue 35, as the Gemtech Integra integrally suppressed rifle uses the same technology.

A bore evacuator effectively delays the movement of the gas, giving it slightly more time before it follows the path of least resistance. The end result is that the majority of the extra gas goes through the muzzle end rather than back-flowing through the barrel and needlessly increasing pressure, gas in your face, felt recoil, and wear.

Dead Air first followed up the Nomad with the performance-enhancing E-brake, and then after some skunkworks testing, they released the Nomad-L at SHOT Show 2020. What you see here is the Nomad-Ti, an all-titanium version of the Nomad-30 that weighs in at a mere 9.6 ounces, including the bundled direct-thread mount.

It's easy to see why Dead Air keeps investing in the Nomad system. From the day they released them, they can't keep them in stock; it was a great performer, and people quickly took notice.

Unlike most Dead Air cans, the Nomad-Ti does indeed have barrel length restrictions — high pressures produced by short-barreled 5.56 guns will heat and eat titanium for breakfast. As such, the Nomad-Ti is limited to 5.56 barrels of 12.5 inches or above for slow-fire.

In addition to being better suited for a 16-inch rifle like the ATAC ADER, it comes naked and uncoated, so it really pops next to the tungsten Cerakote. Remember: Looking cool is always the first rule.

Instead of using the provided Dead Air direct-thread mount, we went with a JMAC Customs X-37 brake. It's compatible with all 1.375x24 TPI silencers, as well as their BDS-37 blast shield for unsuppressed use, should we find ourselves in a state with oppressive NFA restrictions.

OUTFITTING

We attached a US Optics TS-8X first focal plane 1-8x24mm with a

ATAC DEFENSE ADER

CALIBER: 5.56MM

OVERALL LENGTH: 31.5 INCHES

WEIGHT UNLOADED: 6 POUNDS, 4 OUNCES

BARREL LENGTH: 16 INCHES

MAGAZINE CAPACITY: 20, 30

MSRP: $1,399

URL: ATACDEFENSE.COM

DEAD AIR NOMAD-TI

CALIBER: .30

OVERALL LENGTH: 6.5 INCHES

WEIGHT: 9.6 OUNCES

MSRP: $1,099

URL: DEADAIRSILENCERS.COM

FEATURED ACCESSORIES:
SUREFIRE M600DF $299
SUREFIRE SR-07 SWITCH $112
US OPTICS TS-8X PLUMB $895
DUECK DEFENSE RTS SIGHTS $200
JMAC CUSTOMS X-37 $120
ARISAKA FINGER STOP $28
MAGPUL MVG $23

PRICE AS FEATURED: $4,175

5.56mm Plumb Precision reticle (see RECOIL Issue 49 for a review) in a Midwest Industries mount. A set of Dueck Defense offset iron sights served as backups, and a 1,500-lumen SureFire M600DF riding on an Arisaka inline scout mount with an SR-07 remote switch made for a helluva light beacon out front.

We then added a Magpul MVG to use in conjunction with an Arisaka finger stop to ensure consistent hand placement. Some may consider that a suspenders-and-belt setup — but our pants certainly aren't falling down.

AT THE RANGE

The trigger was exceptional; you can go fast as needed but also utilize it for precision work. The recoil impulse unsuppressed is fast but gentle — an adjustable gas block for the ADER is in our near-future for suppressed use; it was workable but not optimal. Though the Nomad-Ti is low-pressure, we'd still like a bit more control of the increased gas pressure. For testing, we rolled with basic-bitch 55-grain M193 ammunition, which turned in perfectly acceptable 1.5-inch five-shot groups. Undoubtedly, match ammunition would result in tighter groups, but this essentially is a fighting rifle, not a DMR.

Hearing safe? Welp, we didn't even consider adding ear protection when shooting outdoors; indoors is a different story with 5.56mm of any stripe or barrel length.

LOOSE ROUNDS

Dead Air Silencers continues to impress us with their innovation. Even though the Nomad-Ti is a straightforward extension of an existing product line, a sub 10-ounce suppressor with this level of performance is exceptional.

While the ADER is the first rifle ATAC Defense has produced, it's far from a freshman effort. Over the last eight years, they've clearly learned what right looks like and put it all out there. While there's some room for changes and improvements, dollar-for-dollar the ATAC Defense ADER is more than a solid choice. R

DOWNSIZING

Sometimes Good Things Do Come in Small Packages

By Alexander Crown

We all know how it goes — you find that roll pin on the floor and now you have to build another gun. In this case, it was being gifted a really nice 9mm AR barrel that pushed everything over the edge to put this together. This build is something that had been pondering for considerable time. Now being "forced" to take action, some goals were identified: a compact 9mm, runs on Glock mags, and can be easily suppressed. This build is meant for fun ground squirrel hunting or occasional plinking when/if ammo prices even out.

COMPONENT	MSRP
Bootleg Inc Enhanced AR-15 Upper Receiver	$180
Spikes Tactical 9mm Spider Stripped Lower	$180
Bootleg Inc M-LOK 4-inch Handguard	$136
Primary Weapons Systems 9mm Bolt	Discontinued
Rainier Arms Select 6-inch 9mm Barrel	$150
Maxim Defense CQB Brace	$190
CMMG Lower Parks Kit	$69
Magpul MOE-K Grip	$19
Bravo Company USA Charging Handle	$40
EXTRAS	
SilencerCo Omega 9K	$865
Aimpoint H-1	Discontinued
LaRue Tactical LT660	$107
SureFire M600 Scout Light	$297
Magpul M-LOK Offset Light/Optic Mount	$35
KRISS MagEx2 Extension	$30
Glock 17 Magazine	$38
Edgar Sherman M81 Sling	$50
X2 Magpul Quick Detach Swivels	$30
Impact Weapon Components 2-to-1 TriGlide	$20
TOTAL	**$2,626**

The Rainier Arms Select Series Barrel was the perfect starting point for this little blaster. At 6 inches, it wasn't so long that subsonic rounds would go supersonic, since this build is intended to be shot suppressed almost exclusively. The barrel came threaded ½-28mm, making the mating of the SilencerCo Omega 9K easy with a direct thread adapter. The bolt was purchased at a steep discount locally due to the product being discontinued. It's from Primary Weapon Systems and features a removable carrier weight. The charging handle is the standard Bravo Company USA Gunfighter Medium, giving us increased surface area on the latch mechanism to help clear the sling QD swivel.

The Bootleg Enhanced Upper doesn't have a forward assist, which would be unnecessary for a pistol-caliber build, and it saves us a little bit of weight (1.7 ounces) over a standard Mil-spec upper receiver. The handguard is extremely easy to install, given the camming mechanism, and at 4 inches is a great length for the barrel. And, of course, M-LOK lets us mount stuff like the SureFire Scout light with Magpul mount to position the light higher at the 1 o'clock position for thumb activation. One gripe is the lack of M-LOK on the bottom, but we can get past this.

The lower receiver was purchased locally and is a Spikes Tactical "Spider" stripped lower. The lower included takedown pins and has a last round bolt hold-open device installed. A standard CMMG lower parts kit fills in the rest of the guts; they may not be flashy, but the CMMG kit fit and didn't break the bank for quality parts. We'd like to upgrade the trigger to the CMC Single Stage 3.5lb 9mm PCC Flat trigger in the future. The Magpul MOE-K is an incredibly compact and thin grip that adds little in the way of weight and sits flush when a Magpul 21-round magazine or Elite Tactical Systems 22-round magazine is installed. An important consideration for storage of the gun.

To keep the overall length of the build

small and functional, a Maxim Defense CQB PDW brace was selected. These can come with a PCC weight buffer for use in blowback systems where more weight is necessary for reliable function. Once fully extended, the Maxim brace goes from 5.3 to 9.2 inches, providing sufficient length of pull for us for control.

The PDW brace also has quick detach mounts machined into the extension for attaching a sling. Since this package is small enough that a single-point sling would work, we selected the Edgar Sherman Design M81 sling with the use of an Impact Weapon Components 2-to-1 Point TriGlide. This little piece takes a two-point sling with quick detach mounts and allows you to make it into a single-point sling quickly. The sling also has the added benefit of an elastic band sewn in for storing the sling on top of itself for when the firearm is in a case or a bag. Topping off the build is the bombproof Aimpoint H-1 red dot riding in a LaRue Tactical quick detach, lower 1/3 cowitness mount.

The culmination of parts is a handy and capable package that fills the given parameters of being small and using Glock magazines. The benefit of Glock magazines is the ability to tailor the size to your needs, ranging from 10 and up to 50 rounds. This little guy tips the scales at 6.76 pounds, unloaded — not the lightest weight of builds, but this is acceptable given the heavy blowback components. Weight could be shaved replacing the PDW brace (but we'd sacrifice overall length) or using a lighter optic/mount combo. It's all give and take, and the important thing is that we have a fun package that can be upgraded as needed or wanted. **R**

FIXED STOCK & NO SLOP

We Build Up from a KE Arms KP-15 Base

By Dave Merrill

The major standout of this build is the complete KP-15 SLT/Ambi polymer receiver from KE Arms. Featuring an integral A1-length buttstock, the KP-15 addresses a common weak point with other synthetic receivers — namely, breaking right where you attach your receiver extension tube and buttstock. While it may seem strange to rock a fixed stock in absence of a federal assault-weapon ban, the A1 length is fairly universal, and the taller among us can easily pop on a buttpad.

The KP-15 is also featured in the InRange/Brownells What Would Stoner Do (WWSD 2020) project. You may have seen a similar receiver from the now-defunct Calvary Arms, but the concept of an all-in-one polymer lower started with Eugene Stoner himself. While Stoner didn't find it viable at the time, we've had more than a few advances in materials science in the intervening decades.

Since this came as a complete lower, the internals are all KE Arms, with the excellent SLT-1 trigger included. The Sear Link Technology trigger maximizes energy transfer with the removal of the disconnector, allowing the selector to be placed on safe regardless of the hammer position, with a crisp break followed by a tactile, short reset.

The upper receiver is standard forged and came from a box o' parts. There's nothing special to see here, and the origins are unknown. The 4150 CMV, black-nitride, 16-inch, mid-length 5.56x45mm NATO 1/7 barrel was provided by DRG. Barrels are one of the categories where we sometimes have to scrimp and scrape during leaner times, so we were more than pleased to get our hands on a quality barrel.

The bolt carrier group is an AXTS (now known as Radian) black nitride model, with a hardened HPT/MPI 9310 steel bolt. While original specs call for bolts machined from Carpenter 158, 9310

and S7 steels have proven to be just as durable in this application and are often easier to source.

While Griffin Armament is best known for their silencers, they have an entire line of AR small parts and furniture. For this build, we used the latest M-LOK Low-Pro RIGID rail — the longest they had available to maximize our real estate with a 16-inch barrel. We kept attachments pretty slick, only adding an Emissary Development M-LOK Handbrake. The Handbrake is larger than a finger-stop or index point but doesn't quite have the girth of a stubby VFG.

The silencer is a Gemtech Trek II. While it won't knock anyone's socks off with some new, yet unattainable level of sound reduction, the Trek II remains a relative short, solid performer with direct-thread attachment and corresponding wrench flats.

The optic mount is a Strike Industries Adjustable Scope Mount (ASM), and there are a few features we'd like to point out. The ASM can be set for one of four different eye reliefs to maximize rifle and scope compatibility, and ours features a preproduction ring top-half replacement, allowing for red-dot mounting. A universal MRDS mounting base can be attached at 90 or 45 degrees and can accommodate most footprints. And if you flip the mount over, you can attach an Aimpoint Micro.

Tiring a tad from the LPVO game, we upped our magnification game with the Lucid MLX. Featuring Japanese glass, mil/mil adjustment, and a first focal plane optical arrangement so the reticle scales along with the magnification, the MLX checks a lot of boxes at a lower price point. While 18x magnification is more than we'd usually ask for a 5.56 rifle, having that higher end can be useful for lower-probability shots or simply helping a friend check a target. A spare Atibal MRD V3 red dot was mounted forward and at 45 degrees for close-in shooting and as a secondary sighting method.

The only future additions we envision are perhaps a bipod for stability and a mounted LEP flashlight to maximize night-time non-NV long-range shooting. It's not quite a kick-around do-all, but it's a well-balanced shooter biased toward the longer end of the spectrum. **R**

COMPONENT	MSRP
KE Arms KP-15 Complete SLT/Ambi lower	$450
White Label Armory/DRG barrel	$160
Upper receiver	~$105
AXTS/Radian bolt carrier group	$185
Griffin Armament RIGID rail	$210
Strike Industries ASM scope mount	$143
Strike Industries optical plate	TBD
Gemtech Trek-II silencer	$449
Atibal MRD V3 red dot	$289
Lucid Optics MLX 4.5-18x scope	$720
Emissary Development handbrake	$35
TOTAL	**$2,747**

Can You Use Off-the-Shelf Components to Build a Bolt Gun to Cover all the Bases?

By Iain Harrison

Photos by Kenda Lenseigne

THE O

Sometimes, you just have to roll up your sleeves and do things yourself.

If you had to assemble one rifle to cover all your intermediate to long-range needs, what would it be? If you asked this question to 100 gun nuts, you'd probably come away with just as many different answers. But there'd be a few common threads running throughout the discussion. Just like if you posed the conundrum regarding choices for putting together your ultimate AR, there's no right or wrong answer. It's entirely situationally dependent, but it's a great mental exercise and worthy of late-night campfire discussions involving brown liquid and burning plant material.

Living in the desert southwest poses different firearm challenges and needs than you might find in New England, where shots longer than 300 yards are going to be in the minority and the availability of common calibers is a more important factor than wringing out the last bit of advantage when it comes to reduced wind drift. Hunting the Rockies while humping an extra couple of pounds of steel and glass means that you're going to be breathing out of your asshole with greater vigor than if you'd carried the same rifle to a Midwest deer stand. Like we said, it's situationally dependent, and your choices might be entirely different than your buddy's.

Running through a personal decision tree, our one rifle had to check the following boxes:

- The ability to hit multiple, torso-sized targets out to the maximum distance they can be observed in our backyard, which realistically is 1,100 yards or thereabouts
- Able to hit the kill-zone on deer-sized animals out to a self-imposed ethical maximum of 700 yards, which equates to an 8-inch diameter circle and a projectile energy of at least 1,000 ft-lb (though more is definitely better)
- Reliable, detachable magazines for multi-target engagements, or for missing a lot
- Light enough to carry for days at altitude, in an environment where there's 25 percent less O_2 available than at sea level
- Capable of accepting common, accuracy enhancing shooter aids, such as tripods, bipods, barricade stops, rear bags, etc.
- Decent aftermarket support in the event of requiring replacement parts
- Reasonable barrel life of at least 2,000 rounds
- Suppressor-compatible
- Able to be custom fit to personal preferences regarding length of pull and comb height — turning a couple of Allen wrenches to get the gun to conform to our body shape is A-OK
- Future proof, as far as possible

Some of these criteria are antagonistic, and compromises are going to be necessary. For example, the ability to hit multiple targets at

IE GUN

extended ranges is exceedingly difficult with a lightweight rifle, while toting a PRS gun in the mountains is a nonstarter. Weight is currency, and you have to scrimp in some areas in order to spend it in others. Toothpick-diameter barrels are great for saving ounces, but after a couple of shots, groups tend to open up and point of impact shifts are noticeable, even close-in. Lightweight stocks typically lack any adjustability needed to fit the rifle to the shooter, sacrificing comfort and making you contort behind the gun — again, not great for making good hits at distance. When it comes to choosing components for this rifle, some hard choices are going to have to be made.

There's a lot to admire in the Defiance AnTi-X action where every fraction of an ounce has been whittled away without compromising its integrity.

ACTION

The Remington 700 action has been the go-to for many years, whether from Big Green itself or the many clones developed to address its shortcomings. Yes, there are better choices, but aftermarket support isn't as widespread for them and we're at the point where the 700 footprint has been sufficiently refined that it's plenty good enough. Defiance Machine, located in Columbia Falls, Montana, makes some of the best 700 pattern receivers on the market, and their AnTi-X model offers significant weight savings. Why the name? According to Mike Lee, Defiance's head of special projects, they wanted to create an action with all the benefits of titanium, but without the drawbacks. "Titanium isn't an ideal material for a bolt action receiver, as it's prone to galling, particularly at the lugs," explained Lee. "It's also hard on tooling, which makes getting a perfect finish that much more difficult, and there's issues with thread failure due to fatigue."

Instead of using Ti, the AnTi-X action uses good ol' American

steel and has been skeletonized everywhere that isn't subject to stress — even its integral recoil lug has been whittled down to minimize weight — but it still feels smooth when the bolt is cycled. Scope bases are machined as an integral part of the action, which means there's two fewer points of failure, and more weight savings are realized as they're hollowed out also. We opted for a zero-degree cant, as our gunsmith puts in the effort to time the barrel, but 20 MOA bases are an option for anyone who feels they might run out of adjustment on their elevation dial.

An M16-style extractor is standard on the AnTi bolt, which is deeply fluted, both as a weight-saving measure and to give dirt and grit somewhere to go instead of binding up the action. All major bearing surfaces are polished and finished with a nitride surface treatment, making them very slick and impervious to rust.

We opted for a short, rather than long action due to magazine availability. If you want to go long, then five-round mags are pretty much your only option unless you feel like springing for a genuine Accuracy International 10 rounder, which will set you back about 150 bucks a pop and dangles out of the action to unacceptable degree. Think Ron Jeremy doing the breaststroke. The AnTi-X action is cut to accept both AICS and AW pattern mags, and more options is a good thing.

BARREL

If noodle tubes are out, how do we get a light barrel without sacrificing rigidity? Carbon fiber has entered the chat. Although they're not exactly new, there have been some advances in carbon-wrapped barrels since their introduction last millennium, resulting in better resistance to wandering and less point of impact shift during extended strings of fire. We've seen carbon barrels that pattern like shotguns after a couple of magazines, and we've shot others that held tight groups longer than our ammo supply lasted, so they evidently aren't all the same.

Helix 6 was born from the fishing industry, where millions of miles of carbon fiber have been used to create poles since the UK firm of Hardy's made the first carbon rod in 1967, and given the volume of material spun, you'd imagine that particular market would have a pretty good handle on how best to adapt its use to rifle barrels. "We use a combination of longitudinal, axial, and diagonal fibers in our lay up — not just filament winding," explained Jon Beagle from Helix 6. "This means that the barrel harmonics are flatter, making it easier to find a factory load that'll shoot well, and if you handload, then nodes will be wider." According to Beagle, there's considerably less resin in a Helix 6 composite, which leads to greater strength and thermal conductivity. Does all this marketing-speak translate to results in the field? Read on and find out.

STOCK

There's a pretty strict divide between hunting and target shooting when it comes to stocks, and there aren't that many people who'll head to the woods or mountains packing a chassis. You'll find even fewer lightweight hunting stocks on the line at a match. Although there are many advantages in terms of adjustability and accuracy, a chassis' weight penalty usually overcomes any inclination to lash one to a pack, and in a

Helix 6 carbon wrapped barrel is threaded 5/8-24 to accept the AB Suppressors Raptor can. Reflex baffle section is optional, but lowers tone and dB levels.

WEIGHT IS CURRENCY, AND YOU HAVE TO SCRIMP IN SOME AREAS IN ORDER TO SPEND IT IN OTHERS.

Shown in its heavy configuration with buttstock and forend weights, the MDT HNT26 chassis offers plenty of adjustment options, including distance from the shooter's palm to the trigger face. The button above the pistol grip is for the folding mechanism.

shooting sport where mass is used for stability, there's no advantage in giving up its better ergonomics. The NRL Hunter series may change this, however. With an all-up limit of 12 pounds in Factory and Open Light divisions, it could just spur innovation in both stock types.

In order to split the difference between ringing steel and busting lungs, we went with MDT's HNT26 chassis, opting for the model incorporating both a folding stock mechanism and an ARCA rail. Although these features add a few ounces, they're worth the weight expenditure and are valuable additions to the base model, which admittedly, isn't very basic. We like folding stocks a lot, and once you've used one on a hunting rifle, it's a tough feature to give up.

Starting off with a magnesium center section, which carries the folding mechanism, action, and magazine, carbon-fiber bits are bolted onto each end to give you somewhere to stick your paws and face. The forend has plenty of M-LOK slots for mounting accessories, including a Picatinny rail, should you want to run a clip-on night vision device, while the buttstock is adjustable for length of pull and comb height. We wanted the ability to max out the NRL's weight limit, so added a pair of M-LOK weights to our shopping cart, along with a heavier pistol grip and a brass length of pull spacer. Doing so damped out recoil considerably when teamed with our can.

TRIGGER

In terms of quality, the last trigger we got on a factory R700 before Big Green closed its doors in 2019 — how can we put this delicately? — sucked monkey balls. It's hard to wring out any sort of decent performance from a precision rifle when wrestling with an inconsistent, heavy, and gritty pull. So in keeping with our build's potential, we wanted a trigger that was as crisp as a new hundo, broke at around 2 pounds, and allowed us to play with overtravel to match our personal preferences. Although it's geared toward the PRS market, Timney's HIT unit when cranked up to its maximum poundage checks all those boxes, so was an easy choice to make.

OPTIC

If the divide between traditional versus chassis stocks is entrenched, it's nothing compared to the optical split separating first and second focal plane scopes. Hunting scopes — lighter, simpler, and easier to use — have almost always used second focal plane reticles, whereas for military users or anyone playing in the PRS field, it's FFP or bust. Having enough information in the reticle to make an instant correction in the event of a first round miss led us to Horus. And for the scope in which to house it, we wound up at Leupold's door.

The Mk5 3.6-18x44 gives everything you might need to hit long-range targets, including an illuminated, info-rich Tremor reticle, wide magnification range, and excellent optical clarity. It checks the box regarding future-proofing, as unlike dedicated BDC reticles and dials, no matter what ballistic solutions are called for by newer cartridges and projectiles, they're supported, so long as you take the time to learn the system to get the most out of it. The Mk5 also hits our weight goals as although it's not exactly featherlight, at 26 ounces it packs way more features on board than could be dreamed of only a decade ago.

SUPPRESSOR

Choosing a can for this project was one of the toughest decisions we had to make. There are so many great designs on the market right now, and customers, as a whole, are better educated than ever before, nudging manufacturers to keep improving their game. Because the host rifle has a 25-inch barrel, we weren't too concerned about the erosive nature of its powder charge — if this was a can destined for a 10.3-inch M16, we'd want a heavy Inconel blast baffle — but lighter weight materials were higher on the priority list.

Drawing from previous work in the automotive turbocharger field, AB Suppressors baffle designs allow for a fully welded tubeless can, in a variety of lengths and weights. Using a direct-thread end cap mount, their hearing safe, six-baffle Raptor can weighs just under 7 ounces and should you opt to screw on the included reflex blast chamber, you can shed a couple of dB

We needed mags without binder plates in order to accommodate the longer-than-SAAMI-spec handloads. Note the custom serial number.

at the cost of an additional quarter pound. With the reflex chamber in place, it has one of the lowest tones we've encountered.

BIPOD

While the rest of our build's components were being sourced, we were contacted by our friends in Austria. With us being huge fans of the Strasser line of rifles, they wanted to let us know about a side project they'd been working on and offered to send over one of their new bipods, which was at the time only available in Europe. Machined in-house from 7075-T6 aluminum and carbon fiber, its featherweight construction seemed tailor made for this project, while sacrificing nothing in terms of ruggedness or functionality. One of the features we really appreciated in the field was that with the legs at 90 degrees, the bipod flexes to allow the shooter to pan left or right to track a moving target. Definitely a quality piece of kit.

CALIBER

This question usually provokes the most heated argument, and it could be argued that it's the least important factor, so we left it till last. There are plenty of short-action calibers, which will fulfill our criteria, and if we were to just press the easy button, then a 6.5 Creedmoor loaded with factory 142-grain ELD-X loads would just about suffice. But what if we wanted more? Elk can be tough critters, and African deer-sized animals are notorious for their unwillingness to shuffle off this mortal coil. 6.5 PRC would be an obvious upgrade but would require a magnum bolt face to accommodate its fatter case head, ruling out any option down the road of simply rechambering in the multitude of 0.473-inch diameter cartridges. While we're big fans of the PRC, we're also aware of reports where users have roached barrels at the 1,000-round mark — further evidence if any were needed that there's no such thing as a free lunch. And with advancements such as .277 Fury and 6.8 TV coming down the pipeline in the foreseeable future, this didn't seem like such a great idea from the perspective of future proofing the build.

Is there anything here and now which would allow the use of a standard bolt face in a short action

Strasser BONE bipod is the lightest, full-featured model available and allows 20 degrees of pan.

308 Win, right, 284 Shehane, left. More powder, higher velocities and better BCs — what's not to like? Well, apart from the whole, no factory ammo thing ...

and yet still give us a leg up over more mundane calibers? We're glad you asked.

The .284 Winchester was introduced in 1963 and was a classic case of the right cartridge in the wrong gun. Had Winchester introduced it in a short-action Model 70 bolt gun, it would've lit up the shooting world like a downed power line in a California forest, but they didn't have a short-action Model 70, so instead dropped it into the mediocre Model 100 semiauto and an odd-looking lever action, which split the difference between tradition and modernity, finding favor with no one. Because it was fielded in crap rifles, the cartridge itself was downloaded way under its potential, so the gun-buying public greeted it with a yawn.

After wandering in the wilderness for decades, it's found favor with F-Class shooters as a well-balanced 1,000-yard cartridge when loaded with 175-grain bullets and shot out of long-action target rifles. While we can't load it to the same OAL in our project rifle, we can use the internal dimensions of AICS mags to load it to 2.95 inches, which leaves enough room for powder to make it more worthwhile. And besides, who doesn't like a bit of cartridge nerdery?

Taking the geek factor to the next level involves blowing out the case walls to gain about 3 grains of capacity while maintaining its 35-degree shoulder, in order to form the 284 Shehane. Normally, we'd steer clear of wildcats but as we have to handload anyway in order to get any kind of worthwhile performance and can still use the parent cartridge without issue, we figured what the hell and ordered a reamer from Dave Kiff at PTG, who moved heaven and earth to get us it on time.

Our barrel was chambered and head-spaced by gunsmith and RECOIL contributor John Brooks, and everything was bolted together before being shaken out in the Arizona desert. The final load recipe involved Lapua cases, Vihtavuori N165 powder, and a Berger 168-grain Classic Hunter, which were assembled after fire forming with a low-end load of A4350 under a 120-grain Sierra soft point. While we can't divulge the final charge weight due to it being off the charts, it did turn in a muzzle velocity of 2,920 fps and a five-shot group size of 0.75 inch. At 300 yards.

Ready to hunt and in its lightest configuration, our rifle weighs a hair over 8 pounds, all up. Adding the can in its reflex setup, along with a heavier pistol grip, bipod, M-LOK buttstock, and forend weights brings its total mass up to 11.5 pounds, ready to ring steel out to 1,400 yards.

Assembling this project was admittedly a pain in the dick. When you can walk into a gun store and find any number of superb bolt guns right off the shelf available in the amount of time it takes to fill out a 4473, why bother going to the effort of piecing together something that's admittedly only marginally better?

Like building race cars or bikes, the last bit of performance is always the most expensive in terms of both time and dollars. We think we've assembled the cream of the crop in terms of what's currently available in mainstream, off-the-shelf components, without getting into the realm of F1 or MotoGP-level, one-off prototypes. It's a rifle to check all the boxes we came up with, and it exceeds those criteria in a number of places. Job done. R

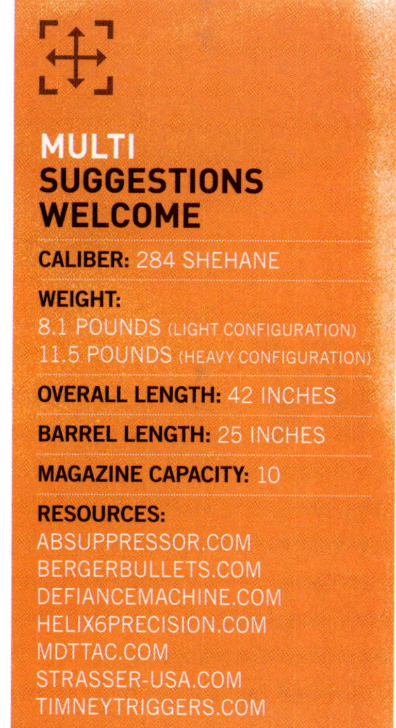

MULTI SUGGESTIONS WELCOME

CALIBER: 284 SHEHANE

WEIGHT:
8.1 POUNDS (LIGHT CONFIGURATION)
11.5 POUNDS (HEAVY CONFIGURATION)

OVERALL LENGTH: 42 INCHES

BARREL LENGTH: 25 INCHES

MAGAZINE CAPACITY: 10

RESOURCES:
ABSUPPRESSOR.COM
BERGERBULLETS.COM
DEFIANCEMACHINE.COM
HELIX6PRECISION.COM
MDTTAC.COM
STRASSER-USA.COM
TIMNEYTRIGGERS.COM

Comprising some of the best components we could find to fit our build criteria, we're pretty happy with the result. It's a beaut, Clark.

MULTI-FUNCTIONAL FRANKENBUILDING

Balancing Flash & Practicality

By Dave Merrill

We took the idea of Frankenbuilding to the next level for this build, as there are scant few matching manufacturers among the parts and pieces used. Ultimately what we ended up with was a rifle that looks a little unusual — a 13.7-inch barrel is rarely paired with a precision stock like the B5. But fear not! There's a method to this madness.

3rd Gen Tactical has been selling precision-machined receiver sets for some time now, and the upper you see here was originally part of a matched Humboldt set. We liked the stylization and subtle lightening cuts on the Humboldt upper and thought it would be fun to match with a different lower. SilencerCo SCO-15 lowers are about more than just the lasered brand name; the major standout is its right-side bolt lock/release. Simply press up on the lever with your trigger finger while racking to lock the bolt and simply press down to release. In addition to modern controls, the SilencerCo lower also has a nice magwell flare, integral trigger guard, and two integral QD cups for slings.

It's been said that the heart and soul of any build is the combination of the bolt carrier group and barrel. The truth of that idiom is certainly up for debate, but we went with high quality anyway. The BCG is from White Label Armory, the retail end of DRG manufacturing, and is paired with a stainless, 1/7 twist, 13.7-inch Noveske barrel. Since this won't be a legally registered short-barreled rifle, extra care was taken with the muzzle device. The Sons of Liberty Gun Works NOX device was specifically designed to accommodate the Dead Air silencers' mounting system and also bring a 13.7-inch barrel up to legal length.

We used a KeyMo adapter for the Witt Machine Mod-1 QD Ultra Compact Suppressor, but feel the OEM mount is superior for this silencer. The Mod-1 works better

The AEMS is chock full of features, and, in particular, we love the crispness of the reticle.

than a silencer of this size deserves to, and it's especially nice on longer barrels. Here, with its stubby length, it's a perfect fit.

The heavier B5 stock on the rear balances the rifle to be more tail-heavy, helping to prevent overswing during target transitions and centering the balance when a silencer hangs from the end.

And we paired it with a Wide Open Trigger. Yes, the hard reset go-faster trigger subject to much controversy. The precision stock says formal, but the WOT says party.

We used leftover Ascend Armory parts for the castle nut, selector, and endplate. Their endplate has a set screw for extra mounting strength and also provides an additional sling QD cup. The short-throw selector has the ability to swap lever lengths and sizes; we kept the right side slick to avoid scraping skin off trigger fingers during manipulation.

The enlarged magazine release is from Battle Arms Development, and those with small fingers will really come to appreciate it. The grip? A surplus A2 with the nubbin shaved off.

The charging handle is a Radian Raptor, arguably the most popular aftermarket charging handle in a decade. Durable, not too large, and actually fully ambidextrous. An Emissary Development Handbrake serves as the forward grip — you can use it like a handstop or pile more of your mitts on it, and it's quickly becoming a favorite among staff.

Covering the barrel assembly is a Troy BattleRail. Slimline and M-Lok, this Troy rail uses a standard barrel nut. A Crimson Trace CMR-301 light/laser combination is on the end of the rail. The laser is powerful and can be picked up in the bright Arizona sun, but the remote switch should be tossed in the bin. Stick to rear-button activation here.

Strike Industries Sidewinder II sights serve as backups, though we're starting to question exactly how many sighting systems a rifle needs. These are ambidextrous with a considerably small profile; they're offset and out of the way until they're needed.

This is the new Holosun AEMS we reviewed in our last issue. It's compact and light at under 4 ounces, with a generous window and crisp 65-MOA circle dot reticle (in red or green) that you can easily use for hosing as well as ranging at distance. Battery life is rated at 50,000 hours, supplemented by a solar panel and shake-awake to save power. Given its circle dot reticle, we felt a compulsion to pair it with an EOTech G45 5x side flip magnifier, allowing it to stretch out its legs a bit.

Precision parts mixed with some flash, along with some svelte and short parts, plus a brand-new optic from Holosun — this rifle is fun as hell to shoot. **R**

PARTS LIST	MSRP
CMR-301: Crimson Trace **crimsontrace.com**	$340
Sidewinder II sights: Strike Industries **strikeindustries.com**	$187
BattleRail 13-inch: Troy **troyind.com**	$188
13.7-inch Infidel 5.56 w/gas block: Noveske **noveske.com**	$430
Precision Stock: B5 **b5systems.com**	$230
Short-Throw Selector, Castle Nut, Endplate: Ascend Armory	n/a
Raptor Charging Handle, Clear Anodized: Radian **radianweapons.com**	$100
SCO-15: SilencerCo **silencerco.com**	$249
Humboldt Upper Receiver: 3rd Gen Tactical **3rdgentactical.com**	$249
Mod-1 QD UltraCompact Silencer: Witt Machine **wittmachine.net**	$499
HandBrake M-Lok: Emissary Development **memissarydevelopment.com**	$35
NOX 5.56 Muzzle Device: Sons of Liberty Gun Works **sonsoflibertygw.com**	$109
KeyMo Muzzle Adapter: Dead Air Silencers **deadairsilencers.com**	$249
G45 Magnifier: EOTech **eotechinc.com**	$699
WOT Hard Reset Trigger: Wide Open Triggers **wideopentriggers.com**	$350
AEMS Red: Holosun **holosun.com**	$471
TOTAL:	**$4,385**

The short Witt Machine Mod-1 QD strikes a great balance of size and performance, especially on a 13.7-inch barrel.

"THE OG"

RECOIL OFFGRID Editor Tom Marshall's Do-All Survival SBR

By Patrick McCarthy

It's been said before that "the most dangerous weapon in the world is a Marine and his rifle." While I wasn't a Marine, I'd like to think that a capable rifleman and his rifle are a force to be reckoned with, Marine or otherwise. As such, a well-tuned and highly capable carbine is an absolute staple of prepared living. There are many like it, but this one is mine.

I started with the military's "recce rifle" concept (essentially an accurized AR), scaled down to leverage current advancements in barrel technology. This build sports a tube from Rosco Manufacturing. Specifically, this one is their collaboration barrel, built with input from Aaron Cowan of Sage Dynamics. It's a 12.5-inch barrel made from 416R stainless with a 1:7 twist and NATO-spec 5.56mm chamber. The barrel is finished in black nitride and features a proprietary "patrol length" gas system that is somewhere between carbine and mid-length. The idea behind this was to create a system more reliable in the short barrel than true mid-length, but without the over-gassing that occurs in carbine-length SBRs — particularly when running suppressed.

This barrel is mated to a lightweight Balios Lite upper from 2A Armament, which is slicked down and lightly skeletonized to reduce weight. The lower is an ADM UIC lower, built for hard-use and sporting a full suite of completely ambi controls. The back end is built around a Primary Weapons Systems enhanced buffer tube and ratcheting castle nut with B5 SOPMOD Bravo stock. The recoil system is a Sprinco "blue" spring with Spikes ST-T1 buffer. A Rise Armament RA-535 trigger, with 3.5-pound single-stage break and a blazing-fast reset, allows for both long-range precision and rapid-fire when needed close-in.

The handguard is an SLR Rifleworks ION Ultra Lite 11.7-inch tube. The ION Ultra Lite features seven sides of M-LOK slots as well as several more M-LOK cutouts along the 12-o-clock axis, with a small section of Picatinny rail at the muzzle end. I tried to make best use of this configuration by clamping a Holosun LS-321 IR laser/illuminator to the front rail section, which is controlled by a Unity Tactical TAPS switch attached directly to the 12-o-clock MLOK slots. The TAPS is a dual-lead switch that single-handedly operates both the laser and white light – in this case, an Arisaka 18650 light body with SureFire UE-series tailcap and Malkoff Devices E2XTL head. This head produces 500 lumens and 55,000 lux. The beam pattern is tight and well-directed with minimal spill, making it especially useful at extended ranges while still providing a large bright spot in CQB. The light is mounted via an Arson Machine inline MLOK mount that pushes the light out almost to the end of the suppressor to minimize barrel shadow. The 12-o-clock position of the TAPS, combined with fully

ambi controls including Radian safety and charging handle means the manual-of-arms is perfectly mirrored regardless of which shoulder you're shooting from.

Speaking of line-of-sight, the optics system consists of a Vortex Razor Gen III 1-10x, mounted in an ultralight Scalarworks mount, paired with a Trijicon RMR on an Arisaka offset mount, which can be configured to either 35- or 45-degree offset to accommodate scope clearance and shooter preference. It should be noted that the Razor is mounted in a 1.57-inch height mount for better cheek weld during prone or long-range shooting, while the RMR is in a 1.93-inch mount for faster acquisition in heads-up shooting, and to permit passive aiming through NVGs.

Furniture is rounded out with a B5 systems Type 23 pistol grip and Strike Industries short angled vertical grip. The Type 23 offers a more vertical grip angle and comes with stippling on both sides and the front strap. We had P4 Coatings add matching stippling to the backstrap for better grip during one-handed manipulations, particularly when running the gun muzzle-up. The Strike short angled vertical grip offers just-enough vertical grip to drive the gun and provide consistent support-hand indexing without excessive protrusion. This grip also has molded-in channels for cable management, but we found Strike's Siegen MLOK inserts to be better suited for our setup. Support-hand grip surface is augmented with insert panels from Walker Defense Research. These polymer panels are dusted with silicon carbide grit to provide a no-fail grip surface while still being lightweight and low profile.

The build is topped off with a Blue Force Gear VCAS sling stowed with a Neomag Sentry Strap. The Sentry Strap wraps around the handguard and is held in place with Velcro, while a small tail with sewn-in rare earth magnets keeps the wrapped up while not in use. To deploy the sling, simply tug on it to break the seal between the magnets and you're ready to go. The barrel is capped with a Rex Silentium MOD X suppressor. The MOD X is a two-part can unique in that, when you order from a Rex dealer, it can be custom-configured to any combination of baffles for both base unit and extension. The base unit can be as short as 3 baffle, and the total baffle count can go as high as 14. We chose to set ours up with a 3-baffle base unit (seen here) with a 5-baffle extension. The base unit is a compact 4.1 inches long and weighs only 8 ounces. The stubby base unit, when used alone, is far from Hollywood quiet but provides enough sound suppression to save your hearing while still being short enough to carry the complete rifle in most SBR backpacks. For more deliberate shooting sessions like classes and range days, the additional 5-baffle unit can be threaded on for better sound suppression.

The sum total of these parts is a single rifle that's capable of both CQB and long-range precision (we've run this rifle past 800 yards with repeatable hits on steel), day and night capable using white light or night vision that can still be packed away in a duffle bag, backpack or between the seats of an SUV. It's truly a go-anywhere, do-anything rifle we expect would serve us well in any emergency. **R**

SELECTED PARTS LIST
ADM UIC-15 Lower ($350) www.admmfg.com
Primary Weapons Systems Enhanced Buffer Tube Kit ($110) www.primaryweapons.com
Rise Armament RA-535 trigger ($260) www.risearmament.com
B5 Systems SOPMOD Bravo ($60) www.b5systems.com
Rosco 12.5" Sage Dynamics K9 Barrel ($170) www.roscomfg.com
Vortex Razor Gen III 1-10x ($3,600) www.vortexoptics.com
Scalarworks LEAP ($399) www.scalarworks.com
Unity Tactical TAPS ($175) www.unitytactical.com
Arson Machine Inline Scout Mount ($48) www.arsonmachine.com
Arisaka 18650 w/Malkoff Devices E2XTL Head ($266) www.arisakadefense.com
Holosun LS321G ($970) www.holosun.com
Strike Industries Short Angled Vertical Grip ($22) www.strikeindustries.com
Strike Industries Siegen Rail Panels ($15) www.strikeindustries.com
Walker Defense Research NILE MLOK panels ($50) www.walkerdr.com
Rex Silentium MOD X ($1,100) www.rexsilentium.com

VNT

We Make a VSS Vintorez at Home with a Liberty Suppressors Leonidas on a Saturday

By Dave Merrill

HOME-MADE AMERICAN OREZ

The VSS Vintorez used to mostly be discussed among Cold War nerds and wildcat reloaders — until 2007. It wasn't combat footage, respect for the terminal ballistics, or appreciation of the design that first captivated public attention. Instead, the glory goes to the first-person-shooter game *STALKER* developed by GSC Game World and released that year.

At least one company has been showing off an American-made Vintorez for a couple years, but we've yet to see it come to market. We're also sometimes cautious of one-off production when it comes to firearms. Ultimately, we wanted our build to be inspired by the Vintorez rather than an exact clone. Everything is off-the-shelf, all mods were done with hand tools, and our base rifle can go back exactly the way it was in under an hour. Winning all around.

9X39MM

The real VSS Vintorez shoots 9x39mm. This cartridge was conceptualized in the Soviet Union to be a dedicated subsonic, suppressed cartridge with better reliability and more mass and ass than subsonic 7.62x39mm at the time. Though there were several approaches to this silent sniper round over the decades of development (such as a 7.62x39mm projectile shoved into a 7.65x25mm case), the Soviets landed on a 7.62x39mm parent case necked-up from 0.311 inch to 0.365 inch (about 9.3mm) to accommodate much larger projectiles.

How much larger? Surplus 7.62x39 is most commonly found with 122- or 123-grain projectiles, with heavier loads like 154-grain occasionally popping up for sale, but 9x39mm starts at 248-grain and goes all the way to a whopping 278-grain. The reports on the armor-piercing variants are largely impressive, but we won't see those stateside unless a foreign power is shooting them at us.

You can safely look at 300BLK as the SAAMI-spec version of JD Jones' .300 Whisper, which in turn is essentially an American version of the 9x39mm. As opposed to necking up a 7.62x39, it only seemed appropriate that for 300BLK a 5.56mm parent case would be used instead.

WHY NOT 9X39MM?

Several years back Wolf Ammunition began importing 278-grain 9x39mm to the United States. There was one major problem: There weren't any firearms in the USA that

An original VSS Vintorez (above, photo by Plomarkie) compared to our weekend project (below).

The Leonidas sleeves an 8.5-inch barrel with a permanently attached monocore.

were capable of firing it. Even though it was derived from an AK-case, manufacturers found it really didn't feed or run well in AK-type rifles. It seems the Russians built dedicated rifles around 9x39mm (the VSS Vintorez followed by the AS VAL) instead of simply re-barreling an AK for a reason.

In a bizarre twist of fate, it was much simpler to develop a 9x39mm AR-15 than anything else, and Marc Krebs of Krebs Custom produced the first viable stateside firearm we saw. A small handful of other companies also produced rifles, with mixed results. But when it hit steel, you could tell.

Back in 2014, the Obama administration instituted a ban on importation of Russian-manufactured firearms, which expanded in 2017. With inexpensive Russian guns no longer available, the American market would answer with USA-made AKs. Similarly in September 2021, right as ammunition panic purchasing started to settle back to normal levels, the U.S. government halted all new ammo importations from Russia. As with AKs, we soon heard of more American companies producing 7.62x39 and 5.45x39 — the two most commonly used Russian cartridges.

Yes, we watched *Red Dawn* too many times growing up.

Unless someone stateside saddles up to produce in quantity, 9x39mm will continue to be a mere curiosity rather than anything remotely practical — 300BLK is dirt common and cheap by comparison.

THE BUILD

While we're sure a competent machinist with the right tools and free time would do a great job, we wanted this to be a project that could be completed on a long weekend day with off-the-shelf parts. Dremels, hacksaws, and paint would be the preeminent tools, but if you have a mill and a Cerakote setup, you're on easy street.

In our minds, the best use of 300BLK is subsonic-only through a suppressed, short barrel. The Liberty Suppressors Leonidas is an integrally suppressed upper that pairs an 8.5-inch barrel with a permanently attached monocore. The Leonidas only requires a single tax stamp, with no federal permission slips needed to cross state lines. It performs exceptionally well with subsonic 300BLK, and importantly the long outer sleeve of the silencer closely matches the lines of the VSS Vintorez.

The main problem is that the Leonidas looks and acts far too much like a standard AR-15, so we had to make some changes.

In lieu of 3D printing or purchasing airsoft Vintorez parts, we went with existing components for everything. An old-school DPMS free-float tube served as the short handguard seen on the Vintorez.

For the receiver, we decided not to risk anything of real value. We had a homebuilt lower that was just a little wonky (See RECOIL's DIY book for our lessons learned) so it would be the perfect testbed for the plan — to hacksaw the magwell to make it more "AK-ish." Similarly, we installed a short-throw ambi selector with a triangular selector fabbed for the right side. It works in "reverse" (flick-up-to-fire) to a normal AK, but it's really not so bad to snap on safe.

We swapped the included Seekins Precision upper receiver and handguard with an inexpensive Bear Creek Arsenal AK-ish right-side-charging upper receiver. We weren't looking for excellent ergonomics — we're mimicking Soviet design here — so the Bear Creek Arsenal upper

was perfect. The swap itself took a considerable amount of time and creativity, as we had to do it without removing the permanently attached silencer core of the Leonidas. We managed to figure out a way to use the existing barrel nut on the BCA upper, but it wasn't a sure thing. Liberty Suppressors also has a Leonidas conversion service, which we'd suggest for anyone who wants to do this themselves. You send in your barreled upper, and they sleeve/weld/finish from there.

The thumbhole stock from NC Star was simply the cheapest one we could find. It was only $35 online and, if we're being honest, far more solid than a $35 stock should be. A fakelite makeover (see RECOILweb for a DIY) pushed it a notch above.

The right-side, reciprocating charging handle is pivotal to this build.

In order to mimic the 4x PSO-1 without having to find a way to attach piss-poor Russian glass to an AR, we chose the 1-6x Atibal Mirage LPVO. Why? Because it was already gray. We sprayed the mount with "hammered" paint to more closely mimic the PSO. For bonus points, we added a rubber front cover and a goofy UTG eyecup.

The sights we attached to the suppressor tube came from a rando pile of old bolt action sights. They aren't functional and are simply glued to the tube — welding, drilling, or tapping would involve permanent modification to a silencer, so we avoided it.

LOOSE ROUNDS

Is our Vintorez more practical than a stand-alone Liberty Leonidas? Not in the least. The short handguard means you're scrunched up when shooting it. The thumbhole stock makes the safety manipulation slightly more difficult. The extraneous eyecup makes the Atibal Mirage difficult to use.

And have you ever wondered why no one chopped the front of their magwells to look more AK-ish? It's because it totally screws up your reloads. JB Weld is for nothing more than dress up (the rear sight fell off after one range trip), so it's a good thing our iron sights were never meant to be functional. If you want them to stay on, you'll probably have to drill/tap or weld them.

But the "Vintorez at Home" is absolutely worth shooting and showing off to your friends.

If you want the thwack of a really fat round on steel, go with an American big bore like .450 Bushmaster. If you want a practical integrally suppressed rifle? Stick with 300BLK.

AMERICAN VINTOREZ

OAL: 34.75 inches

WEIGHT: 8 pounds, 10 ounces

CALIBER: .300BLK

CAPACITY: 10, 20, 30

BUILD PARTS:
Liberty Suppressors Leonidas: $2,094 (or $638 conversion with customer supplied upper)
DPMS Freefloat Tube: $30
Atibal XP-6 Mirage: $430
UTG eyecup: $16
NC Star VISM AR stock: $52
Bear Creek Arsenal Side Charger: $223
80-percent Lower: $85
KE Arms LPK: $90

PRICE AS CONFIGURED: $3,020

BATTLE RIFLE ROYALE

Large-Caliber, Do-All Rifle Build

By Alexander Crown

Despite what some may say, .308 Winchester is still a viable cartridge for hunting, target shooting, and self-defense. Ammo is [somewhat] available and gives the shooter a vast array of bullets and loads to tailor for specific activities. With this in mind, building up a lightweight-ish battle rifle-type AR Carbine became a project.

THE PARTS

Every good carbine starts with a quality barrel and here opted for the Rainier Arms Match .308. This stainless steel, 1/10 twist 16-inch barrel features an intermediate length gas system. The relatively "short" barrel keeps the carbine compact, while still providing plenty of accuracy, even with 147-grain ball ammunition.

Crowning the barrel is the Sig Sauer 762ti-QD silencer. This larger (than most) diameter silencer provides great sound reduction and doesn't tip the scales too much or weigh down the front end.

The 13-inch Centurion Arms 7.62 Rail shrouds the barrel and provides more than enough M-LOK mounting options, plus built-in sling quick-detach sockets on both sides and at the front and rear. Installation was slightly more labor intensive than other rails but the end result is rock-solid with no flex or movement. Not to mention we think it just looks sexy. A bigger gun needs a brighter light, so the SureFire Dual Fuel Scout Light was chosen for its high output of 1,500 lumens. This is situated on the left side of the gun for activation with the thumb.

An Aero Precision M5 Receiver set is a commonly chosen base for rifles these days due to availability, reasonable price, and overall good quality. Buying the complete lower assembly ensures the use of proper .308 parts such as the buffer tube and buffer, which differs in size and weight compared to standard AR-15 buffers. The Radian Talon Ambidextrous Selector is set to 45 degrees and has become a standard on many of the ARs in the armory. A Magpul MOE-K grip was chosen due to some of us having petite hands. All of the other guts in this lower are from Aero Precision.

Our bolt carrier group is a test and evaluation sample from Bootleg Inc., known for their adjustable bolt carrier groups for use in 5.56 rifles and now

scaled up for larger-caliber ARs. Due to this being a T&E unit, it has no markings but does have four different adjustments to vent gas from the ejection port. Using a silencer will certainly increase the amount of gas coming back into the rifle, and this simple solution prevents the need for further modifications like an adjustable gas block. The BCG can be adjusted while in the rifle with a simple flat head screwdriver or even a piece of brass. At the time of this writing, Bootleg expects the Adjustable BCG to be available in the second quarter of 2022.

The B5 Systems Precision stock is adjustable on the buffer tube via a pull-down lever that also has a tension lock so you can adjust the stock without accidently removing it. Comb height and length of pull adjustments are toolless. You also get two rotation limited quick-detach sling points and two hard mount sling points. The bottom of the stock has two M-LOK mounting points for monopods or other accessories; they aren't really applicable for this build, but it's nice to have them for future use. The stock comes in two different variants for 5.56 and .308, the main difference being the latter has a shorter cheek-piece to accommodate the longer charging handle on .308 receivers. Additionally, the B5 stock only weighs 20.5 ounces.

We debated heavily on what optic would fit this build, and while the culmination of parts could easily be topped off with a higher power magnification optic, we opted for something smaller and lighter. The Nightforce NX8 1-8 was the winner and provided us with a first focal plane optic with near-red-dot-capability and 8x magnification for reaching out a little farther. This NX8 is cradled in an American Defense Manufacturing Recon mount.

The sling is an Edgar Sherman Designs ESD Sling in Multicam Arid, which stores on itself via the use of an elastic band sewn right in. The end of the handguard has a Maxim Defense M-RAX piece for a bipod, and the charging handle is a Radian Raptor Ambidextrous. The Magpul PMAG 25 LR/SR is an excellent option for .308 rifles to get an extra five rounds and increase overall ammunition carrying capacity.

The next few upgrades for this build will be an upgraded trigger, backup iron sights or an offset red dot, and a sweet Krylon paint job. Piecing together a battle rifle in this modern day is a fun undertaking, and now that battle rifles of yesteryear like the G3, H&K91, and FAL are getting harder to find and increasingly expensive, this is a great option for those yearning for larger-caliber fighting guns. Parts are becoming widely available again, and the options make it a buyer's playground. **R**

PART	PRICE
Aero M5 Upper Receiver	$150
Rainier Arms Match 16-Inch Barrel	$315
Sig Sauer Taper-Lok Muzzle Brake	$100
Bootleg Inc. Gas Block	$30
Bootleg Inc. Gas Tube	$16
Centurion Arms 7.62 13-Inch M-LOK Rail	$245
Radian Raptor Charging Handle	$90
Bootleg Adj. Bolt Carrier Group	$309
Aero M5 Lower Receiver Complete	$365
Radian Ambi Talon Selector	$55
Magpul MOE K Grip	$21
B5 Systems Precision stock	$226
Nightforce NX8 1-8	$1,750
American Defense 1.93-in QD Mount	$240
SIG 762ti-QD Silencer	$1,150
Edgar Sherman ESD Sling	$50
Magpul 25 round .308 Mag	$25
SureFire Dual Fuel Scout Light	$329
TOTAL:	**$5,466**

BLUE COLLAR ONE STAMP

By Iain Harrison
Photos by Kenda Lenseigne

Pistol-caliber carbines have come a long way in the past decade, so in this build we wanted to construct something that would serve as a cost-effective training tool, a decent competition gun, and something that was just downright fun to shoot. It's a quintessential parts bin special, pulled together from components we had laying around gathering dust. It wasn't until we'd bolted everything up that the discovery that one of the parts was no longer available came to light.

The core of the build is a KE Arms polymer 9mm lower, equipped with their DMR trigger and 45-degree selector, coupled with a heavy, 9mm-weight buffer. One of the attributes this lower brings to the table (apart from the considerable weight savings over an aluminum version) is its adjustable, fixed ejector, which is capable of accommodating a wide variety of 9mm bolts from various manufacturers. This makes tuning the gun much easier, as unlike a 5.56 rifle, there's no Mil-spec or TDP to work to. Based around the ubiquitous Glock mags, the lower has an ambi mag release and selector, as well as a magwell that Stevie Wonder could find. In keeping with the cost-effective theme of this build, the KP9 is considerably less expensive than a typical billet aluminum lower receiver, especially when you add

in the cost of a decent trigger, pistol grip, stock, and receiver extension.

A 7.5-inch Odin Works 9mm barrel was sourced from Primary Arms, as was an Odin Works slickside upper receiver. We permanently attached a Liberty Suppressors Mystic X can to it, bringing the whole unit up to a legal 16.1 inches, which fits conveniently under an SD handguard from Icarus Precision. This handguard is no longer available, but they've indicated that if enough people pester them, they'll bring it back — it really is a well-designed and nicely machined unit that'll accommodate cans up to 1.5 inches OD.

For an optic, we selected a Holosun AEMS reflex red dot, which at almost 500 bucks is definitely stretching the budget, but worth it for the versatility it offers. The entire package weighs in at a very svelte 6 pounds, 6 ounces and points like a magic wand, balancing just behind the front pivot pin. Shot at a local steel match, the KP9-based 9mm AR turned in a creditable performance against much more expensive PCCs, proving that its blue-collar design ethos doesn't hold it back. **R**

PART	PRICE
KE Arms KP-9 polymer lower receiver with DMR trigger and 45-degree ambi selector	$500
Odin Works 9mm barrel	$154
Odin Works 9mm slick side upper	$129
Faxon 9mm bolt	$165
Icarus Precision SD handguard	$N/A
Liberty Mystic X suppressor	$699
Holosun AEMS holographic sight	$471
TOTAL:	**$2,118**

» SOURCE

faxonfirearms.com
holosun.com
kearms.com
icarusprecision.com
libertycans.net
odinworks.com
primaryarms.com

4

POLITICS

ORIGINS OF THE NFA

Making and Apprehending Criminals Through the Tax Laws

By Alexandria Kincaid

A hand-processed, paper-intensive, firearm background check system that costs Americans millions of dollars annually to implement is no doubt an archaic way of doing business in the year 2017. But few, if any, people have ever suggested that government processes are the model of efficiency. Unfortunately, while change may be on the horizon, it'll take much effort and many years to see improvement for the simple reason that far too many people still agree with the following statement:

> "A sawed-off shotgun is one of the most dangerous and deadly weapons. A machine gun, of course, ought never to be in the hands of any private individual. There is not the slightest excuse for it, not the least in the world, and we must, if we are going to be successful in this effort to suppress crime in America, take these machine guns out of the hands of the criminal class." Testimony of Attorney General Homer Stille Cummings as recorded in *National Firearms Act: Hearings on H.R. 9066 Before the H. Comm. On Ways & Means*, 73rd Cong 1 (1934) [NFA Hearing].

"Predatory criminals." "A very serious national emergency." "The armed underworld." These dramatic emotion-provoking descriptions uttered by Attorney General Cummings in 1934 during the first few minutes of his Congressional testimony were designed to sway Congress into passing national gun control through America's tax code. As the spokesperson for the Department of Justice, his focus on addressing crime by restricting and inconveniencing law-abiding Americans hardly differed from the gun-control rhetoric disgorged by today's anti-gun politicians. In his effort to "sell it," Cummings blatantly declared that law-abiding Americans needed to endure the inconvenience of the law to allow the government to deal with criminals.

Without any substantiation, Cummings told his audience that twice as many people existed in the armed underworld as there were in the Army and the Navy combined. His proposed gun-control law imposed a mostly unaffordable tax on importers and manufacturers of firearms, dealers, machine guns, and most other firearms. Firearms excluded from the law were "ordinary shotguns or rifles." It also created a national registration system that would easily allow the arrest, prosecution, and conviction of anyone in possession of a firearm without the proper paperwork.

One difference between Cummings and modern gun-control advocates is that Cummings admitted that a ban on any particular firearms, including machine guns, would be unconstitutional. But, he claimed, Congress could tax firearms instead. He urged Congress to impose a 100-percent tax on machine guns. And that is, of course, ultimately what they did.

The National Firearms Act (NFA) imposes a $200 tax on machine guns and other firearms (with a $5 tax on items classified as "any other weapons"). In 1934, when the NFA passed, this $200 tax was equal to the average cost of a machine gun. To set the stage for how much an imposition the tax was on Americans, keep in mind that according to the Bureau of Labor Statistics, in 1934, the average person earned $1,524 per year. Also for comparison's sake, the average house in 1934 cost $5,970, and a Studebaker truck cost $625. This 100-percent tax was extremely prohibitive for the average American, equivalent to over $3,600 today, according to the U.S. Inflation Calculator.

But what caused the most concern from this initial draft of the NFA for Second Amendment supporters wasn't the misplaced blame on sawed-off shotguns and machine guns or the 100-percent tax. The bigger problem faced by Constitution-loving Americans was that the bill proposed a tax on, and a national registry for, the ordinary pistol and revolver.

Due to the rise in crime that Prohibition and the Great Depression created, the laws to deal with their collateral damage quickly followed. Although the Volstead Act was repealed in 1933, the NFA was enacted in 1934.

The initial NFA bill, H.R. 9066, defined "firearm" as "a pistol, revolver, shotgun having a barrel less than 16 inches in length, or any other firearm capable of being concealed on the person, a muffler or silencer therefor, or a machine gun."

By including pistols and revolvers, the NFA followed the trend of the time. Gone were the days of the 1700s and 1800s when the law *required* Americans to be armed. In the 1920s, other bills had already been proposed to restrict pistols from crossing state lines. Ironically, the argument in favor of one of these pistol-restricting bills was that homeowners could instead keep a "sawed-off shotgun," which, according to Senator John K. Shields, a Democrat from Tennessee, was "far more deadly and surer than the pistol."

More handgun-restricting bills followed Senator Shields's failed attempt. In 1930, several bills were proposed to restrict interstate commerce in pistols, revolvers, machine guns, and shotguns or rifles that had their barrels sawed off or shortened (without specifying a length). *Firearms: Hearing on H.R. 2569, H.R. 3665, H.R. 6606, H.R. 6607, H.R. 8633, and H.R. 11325 Before a Subcommittee of the H. Comm. On Interstate & Foreign Commerce*, 71st Cong. 1-3, 7 (1930).

Clearly, H.R. 9066's focus on pistols and revolvers was nothing new to Congress. But to date, the prior bills had failed. The old saying that laws are like sausages (no one should see them being made) held true in the makings of the NFA. The NRA and other opponents to the bill ultimately relented by accepting the tax on and registration of firearms, if pistols and revolvers were removed from the bill's mandates. The compromised, restrictive NFA that gun owners live with today does very little, if anything, to deter crime, fails miserably as a revenue raiser, and costs Americans millions of dollars annually to implement.

Keep in mind that the central idea of the initial draft of the NFA was to skirt the Constitution and Congress's inability to enact a federal ban of firearms, and instead use Congress's revenue-raising power and its power to regulate interstate commerce to tax and register almost all firearms (except non-concealable rifles and shotguns). This end-run around the limitation on Congress's power and the Constitution was admitted repeatedly during the hearing.

It's interesting to note that H.R. 9066's original definition of "firearm" failed to include items that ultimately wound up in the NFA as enacted. For example, the earlier bill didn't include silencers for firearms not capable of being concealed on the person, such as for rifles or shotguns. It also didn't include rifles with barrels under a certain length. Unnamed "experts" had purportedly been consulted about what the length of a "sawed-off shotgun" was, and reportedly, the bill was lenient in allowing barrels as short as 16 inches, because 18 or 20 inches was a "better maximum length."

The current law exemplifies the arbitrariness of what constitutes a "sawed-off" length, as it allows rifle barrels of 16 inches, but shotgun barrels must be 18 inches. H.R. 9066 also had what can only be described as a "FUBAR" definition of "machine gun," which included both automatic and semi-automatic rifles if their magazines held 12 or more rounds.

GUN CONTROL THEN AND NOW

Not surprisingly, the authors of the initial gun-control bill were similar to today's usual gun-control suspects — ignorant of firearms, their forms, and their functions. The bill was, like many gun control laws, a feel-good, emotional response on behalf of the ignorant to the crime problems of the time.

In the NFA hearing, Karl T. Federick, President of the NRA, pointed out that the bill's definition of "machine gun" was "wholly inadequate and unsatisfactory." As mentioned, the proposed definition deemed a gun that fires automatically or semi-automatically more than 12 shots to be a machine gun. Frederick pointed out the distinguishing feature of a machine gun was not its magazine size, but its function upon a single pull of the trigger. During his testimony, one frustrated Congressman (Woodruff), in addressing Frederick, said "magazine or the clip or whatever they use to hold these cartridges." If an image of Dianne Feinstein just crossed your mind, take a moment to delete it and read on.

Gangsters like John Dillinger were mentioned by name at the NFA hearings as a method of justifying its necessity.

Despite not having any data, statistics, or even suggestion that these firearms were used in criminal activity (or that the tax would control the criminals), lawmakers advocated to add additional firearms that would be subject to the tax. For example, later in the bill's hearing, Republican Harold Knutson from Minnesota suggested adding rifles with a barrel under 18 inches to the bill. There had been no suggestion that shorter rifles were a crime problem. Similarly, the record is completely devoid of any factual data that suppressors were used in criminal activity warranting any type of restriction, and no one bothered to bring this issue up.

CRIMINALS WILL STILL GET THE GUNS

At the initial NFA Hearing, Adjt. General Milton A. Reckord, who was also vice president of the NRA, got directly to the point that the DOJ wasn't approaching the crime problem properly.

Reckord compared the law to the Volstead Act: "The honest citizens are not going to be bothered with such restrictions. They won't obey the law, and you are going to legislate 15 million sportsmen into criminals; you are going to make criminals of them with the stroke of the President's pen."

Both Major General Reckord and NRA President Frederick pointed out that any knowledgeable person knew that the criminals would still get guns, despite the new law. In fact, the famous gangster, John Dillinger, who was repeatedly mentioned during the hearing as a primary example for why the law was needed, stole his guns. Reckord and Frederick suggested that the correct solution would be for the government to focus on punishing criminals with guns.

The bottom line is that no one, not even the Attorney General, expected the criminals to comply. In agreeing that the criminals would not obey the law, Cummings argued that the bill was designed to make it easy to convict criminals for their noncompliance with the tax code — if a suspect has a machine gun and no paperwork, they could be prose-

cuted. He cited the arrest, prosecution, and conviction of Al Capone as an example of a known, violent, and slippery criminal who was apprehended because of the income-tax law that he violated. The attorney general's goal, and his logic, was that this new tax law would similarly help law enforcement apprehend the John Dillingers of the time for violating the tax code rather than any of their violent criminal activity.

Many of the arguments against the NFA were similar to the arguments made by Second Amendment supporters today: gun control doesn't affect criminals. Automobiles cause more deaths than firearms. Handgun owners with concealed-carry permits are law-abiding citizens, and crime will increase if you take guns away from them. Frederick also made the point that police forces in rural communities are inadequate, and the law would prevent people in smaller communities from obtaining weapons needed for self-defense. Those in favor of the new law replied that people don't need pistols for self-defense, just rifles or shotguns, and the Committee Chairman claimed that he had "never heard" of anyone needing a pistol for self-defense.

Frederick pointedly stated that, "I do not think we should burn down the barn in order to destroy the rats. I am in favor of some more skillful method of getting the rats without destroying the barn." He reiterated that gun-control laws don't reach the "crook" at all, just the honest man. To support his position, and to combat the suggestion that people don't use firearms lawfully for self-protection, Frederick had compiled and provided newspaper articles on law-abiding citizens using firearms in self-defense. His arguments, of course, had factual support, unlike the purely emotional arguments in favor of the bill.

WHAT ABOUT THE SECOND AMENDMENT?

Not everyone was willing to accept an end-run around the Second Amendment to apprehend violent criminals for non-violent behavior. Some even saw the danger of criminalizing law-abiding Americans who simply didn't have the correct paperwork. Rep. David J. Lewis of Maryland commented on what was to him a blatant attempt to violate the Second Amendment. He stated that he "never quite understood how the [gun control] laws of the various States have been reconciled with the provision in our Constitution denying the privilege to the legislature to take away the right to carry arms." Cummings responded to Lewis that, "We are dealing with another power, namely, the power of taxation and of regulation under the interstate commerce clause. You see, if we made a statute absolutely forbidding any human being to have a machine gun, you might say there is some constitutional question involved. But when you say 'We will tax the machine gun' and when you say that 'the absence of a license without payment of the tax has been made indicates that a crime has been perpetrated,' you are easily within the law." Cummings's logic was simple: No prohibition of firearms was proposed, just regulation and criminal prosecution for those who did not comply.

DEALERS AND MANUFACTURERS

The proposed law would not tax only the purchasers (transferees) of the NFA-defined firearms, but the dealers and manufacturers as well. One Congressman put on the record that he would like to put pawn-brokers and dealers in used firearms ("those people," in his words) out of business, if he could. Interestingly, W.B. Ryan, president of Auto Ordnance Co., manufacturer of the Thompson submachine guns, spoke in support of the NFA.

When NRA President Frederick spoke to the issue of taxing dealers, he pointed out that an annual dealer tax of $200 would eliminate 95 percent of the dealers in pistols. Similarly, the proposed $5,000 tax on manufacturers was much too high for any smaller companies to pay.

MODERN NFA

After the initial hearing on H.R. 9066, a modified version of the NFA was proposed in June of 1934, H.B. 9741. This is the bill that was ultimately passed and enacted as the National Firearms Act. Representative Robert Lee Doughton, a Democrat from North Carolina, introduced this bill. It passed the

House and Senate and was signed into law by President Roosevelt all in the same month. It was, as suggested in the final hearing by Rep. Doughton, a law that was supposed to no longer put the citizenry at the "mercy of the gangsters, racketeers, and professional criminals."

He touted that the bill no longer affected pistols and revolvers, so that "law-abiding citizens who feel that a pistol or a revolver is essential in his home for the protection of himself and his family should not be classed with criminals, racketeers, and gangsters, should not be compelled to register his firearms and have his fingerprints taken and be placed in the same class with gangsters, racketeers, and those who are known as criminals." Congressional Record, 73rd Congress, June 13, 1934, 11400.

The NFA, as enacted, defined "firearm" as "a shotgun or rifle having a barrel of less than 18 inches in length, or any other weapon, except a pistol or revolver, from which a shot is discharged by an explosive if such weapon is capable of being concealed on the person, or a machine gun, and includes a muffler or silencer for any firearm whether or not such firearm is included within the foregoing definition." Since 1934, this definition has been revised and expanded through additional laws, such as the Firearm Owners Protection Act, as well as through the ATF's interpretations of the statute (such as the ongoing "Sig Brace" conundrum). Suffice it to say that the NFA's definition of firearm is a living, breathing beast that has not yet been slain.

Merchants and law-abiding citizens continue to bear the burden of outdated, costly, and unnecessary legislation.

Why are these particular firearms subject to the tax code and registration? Because in the eyes of the 73rd Congress, they were the firearms used by the criminals whom law enforcement were somehow unable to apprehend any other way.

Although the current NFA remains a gun control law written in a tax code (found in Title 26, United States Code section 5801 et. seq., which is the Internal Revenue Code and enforced by the Bureau of Alcohol, Tobacco, Firearms, and Explosives, formerly under the Department of the Treasury, but now overseen by the Department of Justice), it carries hefty criminal penalties. As intended, the law allows the apprehension and prosecution of those who don't pay the tax or who otherwise fail to comply with the registration (paperwork) requirements. As predicted by Major General Reckord, it's a pitfall for unsuspecting gun owners who can easily and unknowingly violate the NFA and commit an accidental felony.

While this "tax code" was touted as a law that wouldn't place law-abiding Americans in the same class as criminals, it in fact does. Given our modern technology and the later passage of the Gun Control Act (which requires modern, computerized background checks), we should continue the effort to eliminate this costly and unnecessary piece of legislation. **R**

About the Author

Alex Kincaid is a former elected district attorney and current firearms law attorney with over 19 years experience. She is an active proponent of the Second Amendment, a legal analyst, and author. Alex hosts an online show highlighting current firearms law-related news called *The AK Show*.

www.alexkincaid.com

DEATH AND TAXES

A Walk-Through of the ATF eFile Process

By Tom Marshall

With the focus of this issue being on short-barreled-everything, we figured it only appropriate to spend a little time talking about the legal process to obtain your very own not-so-long gun.

Despite the undisputed sex appeal of SBRs, the extraneous red tape required to get one is more than a little intimidating at times. In a rare stroke of common sense, the ATF has established an electronic system for filing some of their NFA-related forms, including the Form 1 "Application to Make and Register a Firearm." This is the form to use when you are making an SBR yourself. If you're purchasing a factory-configured SBR, you'll need to file a Form 4, which is a separate process. It should be noted that "making" a firearm doesn't require a professional assembly line or QC shop. Slapping a store-bought 10.5-inch AR upper on a lower you already own counts as "making" a firearm. In this case, you'd simply file a Form 1 to get the lower approved to take shorty uppers.

The screenshots you see here were for the registration of the Project JSOK short AK seen elsewhere in this issue. In that case, we obtained a receiver from Petronov Armament on a standard Form 4473. Then, we e-filed a Form 1 for that receiver before building it out. The e-file process itself is surprisingly user-friendly. The website does run a little slow and sometimes requires multiple clicks on something to get it to work. But it does run. Here's what we had to do to get our Form 1 into the pipeline:

1. Create user profile on the ATF eForms portal

Enter your basic contact info like name, address, and phone number. Then, choose a password. Your username is automatically generated by the system. Once all the required fields are filled in and submitted, you'll receive an email from ATF with your information.

2. Log in to eForms and select the appropriate form to file

In this case, we're filing a Form 1, as an individual. There are "special instructions" for individuals, mostly to do with fingerprints.

3. Fill out the digital version of the Form 1

This consists of several pages' worth of information, including:

Eligibility Questions: Just like the 4473, there's a list of questions about your criminal history and citizenship in order to determine eligibility.

List of Responsible Persons: All of the individuals covered by this application. Current policy requires each person being listed to fill out a questionnaire and submit fingerprints. You must also upload

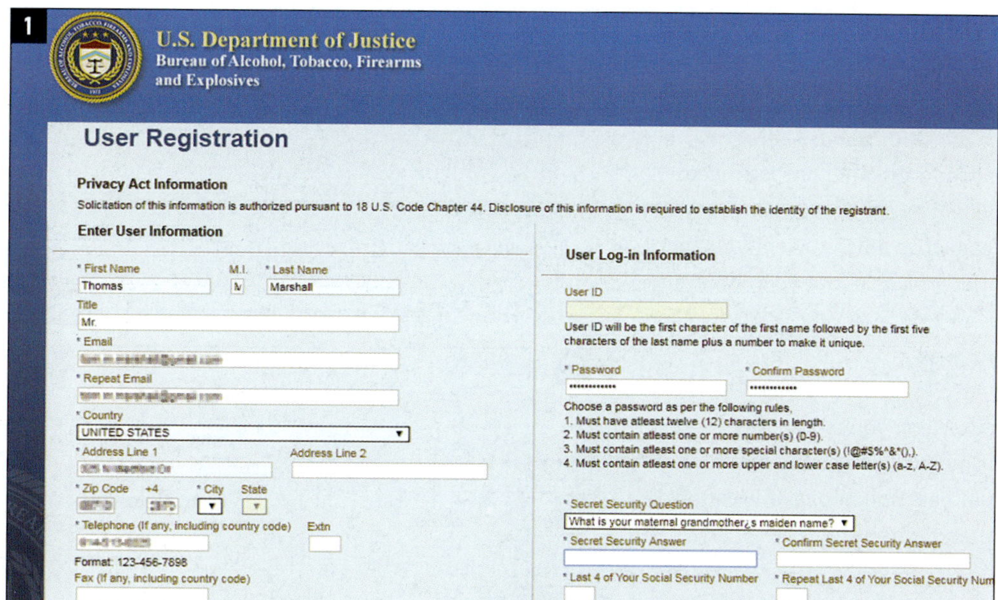
1

U.S. Department of Justice
Bureau of Alcohol, Tobacco, Firearms and Explosives

User Registration

Privacy Act Information

Solicitation of this information is authorized pursuant to 18 U.S. Code Chapter 44. Disclosure of this information is required to establish the identity of the registrant.

Enter User Information

* First Name: Thomas | M.I.: M | * Last Name: Marshall
Title: Mr.
* Email
* Repeat Email
* Country: UNITED STATES
* Address Line 1 | Address Line 2
* Zip Code | +4 | * City | State
* Telephone (If any, including country code) | Extn
Format: 123-456-7898
Fax (If any, including country code)

User Log-in Information

User ID

User ID will be the first character of the first name followed by the first five characters of the last name plus a number to make it unique.

* Password | * Confirm Password

Choose a password as per the following rules,
1. Must have atleast twelve (12) characters in length.
2. Must contain atleast one or more number(s) (0-9).
3. Must contain atleast one or more special character(s) (!@#$%^&*()_).
4. Must contain atleast one or more upper and lower case letter(s) (a-z, A-Z).

* Secret Security Question: What is your maternal grandmother¿s maiden name?
* Secret Security Answer | * Confirm Secret Security Answer
* Last 4 of Your Social Security Number | * Repeat Last 4 of Your Social Security Num

2

FORM 1
ATF F 5320.1
ATF Form 1 (5320.1)
3 of 9

3a

U.S. Department of Justice
Bureau of Alcohol, Tobacco, Firearms and Explosives
eForms

Home | My Profile | FAQ | Ask The Experts

ATF Form 1 (5320.1) - Application to Make and Register a

Applicant Type

* Application is made by *(Select one)*:
Trust | Corporation Or Legal Entity | Individual | Government

Cancel | Next

3b

to the prohibitions on the possession of a machinegun, an application to make a machinegun will generally be disapproved unless the application was submitted by a government agency.

Special Instructions for Individuals, Trusts and Legal Entities

FILING AS AN INDIVIDUAL

The applicant must attach a digital photo in the Photo function on the Responsible Person tab.
The applicant must also provide his or her fingerprints on Form FD-258. The fingerprints will be submitted separately from the application. Upon submission of the application, the eForms system will email a cover sheet to the applicant. The cover sheet is to be printed as it will provide the control number of the transaction and the address where to send the fingerprint cards. The cover sheet and fingerprint cards MUST be mailed to the NFA Division within 10 business days of filing the application.

FILING AS A TRUST OR OTHER LEGAL ENTITY

In addition to providing the documentation of the existence of the trust or other legal entity, the applicant must electronically provide a photograph and completed ATF Form 5320.23, NFA Responsible Person Questionnaire (RPQ), for each responsible person (see definition 1.e on the instruction page) of the trust or legal entity. When filed via eForms, when in the Responsible Person function, the applicant will identify each responsible person and attach the photograph and a scanned, signed RPQ from that person. All photographs and RPQ forms must be attached to the application prior to submission.

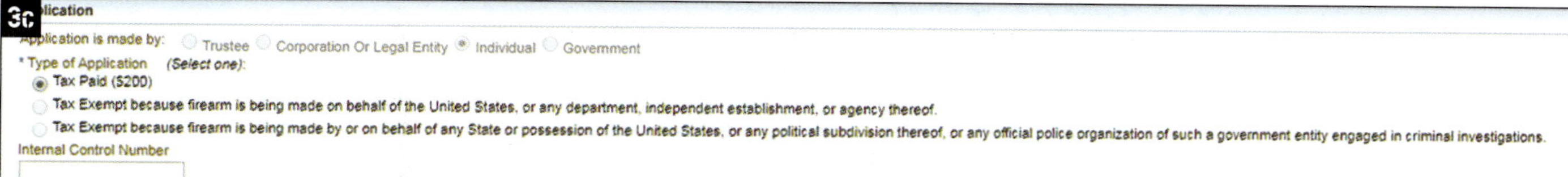
3c

lication

Application is made by: Trustee | Corporation Or Legal Entity | Individual | Government
* Type of Application *(Select one)*:
Tax Paid ($200)
Tax Exempt because firearm is being made on behalf of the United States, or any department, independent establishment, or agency thereof.
Tax Exempt because firearm is being made by or on behalf of any State or possession of the United States, or any political subdivision thereof, or any official police organization of such a government entity engaged in criminal investigations.
Internal Control Number

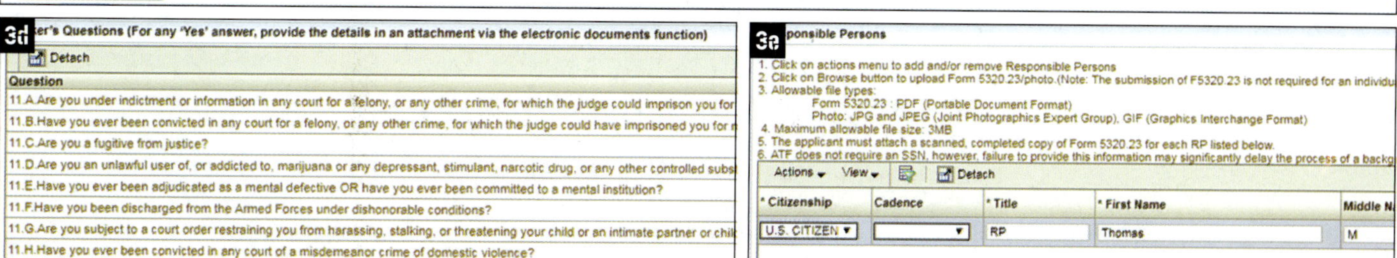
3d

er's Questions (For any 'Yes' answer, provide the details in an attachment via the electronic documents function)

Detach

Question
11.A Are you under indictment or information in any court for a felony, or any other crime, for which the judge could imprison you for
11.B Have you ever been convicted in any court for a felony, or any other crime, for which the judge could have imprisoned you for m
11.C Are you a fugitive from justice?
11.D Are you an unlawful user of, or addicted to, marijuana or any depressant, stimulant, narcotic drug, or any other controlled subs
11.E Have you ever been adjudicated as a mental defective OR have you ever been committed to a mental institution?
11.F Have you been discharged from the Armed Forces under dishonorable conditions?
11.G Are you subject to a court order restraining you from harassing, stalking, or threatening your child or an intimate partner or chil
11.H Have you ever been convicted in any court of a misdemeanor crime of domestic violence?

3e

ponsible Persons

1. Click on actions menu to add and/or remove Responsible Persons
2. Click on Browse button to upload Form 5320.23/photo.(Note: The submission of F5320.23 is not required for an individu
3. Allowable file types:
Form 5320.23 : PDF (Portable Document Format)
Photo: JPG and JPEG (Joint Photographics Expert Group), GIF (Graphics Interchange Format)
4. Maximum allowable file size: 3MB
5. The applicant must attach a scanned, completed copy of Form 5320.23 for each RP listed below.
6. ATF does not require an SSN, however, failure to provide this information may significantly delay the process of a backg

Actions | View | Detach

* Citizenship	Cadence	* Title	* First Name	Middle N
U.S. CITIZEN		RP	Thomas	M

3f

F does not require an SSN, however, failure to provide this information may significantly delay the process of a background check and delay the rendering of a final decision on the application.

tions | View | Detach

SSN	* Zip Code	* State	* City	* Street	* Date of Birth	* Birth Country	Birth State	* State Of Residence	* Sex	* Race	UPIN
		ARIZONA	Tucson			UNITED STA	NEW YORK	ARIZONA	MALE	WHITE	

a photo of yourself to include with the application.

CLEO Notification: There's no longer a requirement for the applicant to get permission from their respective chief law enforcement officer (CLEO), but you do have to send a copy to your CLEO for the purpose of notification.

Line Item: You're required to select or manually enter the make, model, and description of the firearm you intend to register.

Electronic Documents: Any supporting photos or documents.

4. Pay Tax Stamp

Break out your wallet and fork over your allowance money. Fortunately, the eForms systems allows you to pay with credit card, so you don't even have to actually have money to get the ball rolling.

5. Submit Fingerprint Card

Unfortunately, the eForms system cannot accept fingerprint cards. Once you submit your Form 1, the eForms automated system will generate a cover letter for you to mail in with a standard print card. The cover letter includes your application number so that when you mail everything in to NFA Branch, they know who to match your application up to.

6. Wait

The worst part. At time of writing, the author has heard everything from six days to seven months in terms of Form 1 lead times. Once the fee is paid and the paperwork is in, your best bet is to find a few other guns to occupy your time until the ATF can run through your application and grant your tax stamp.

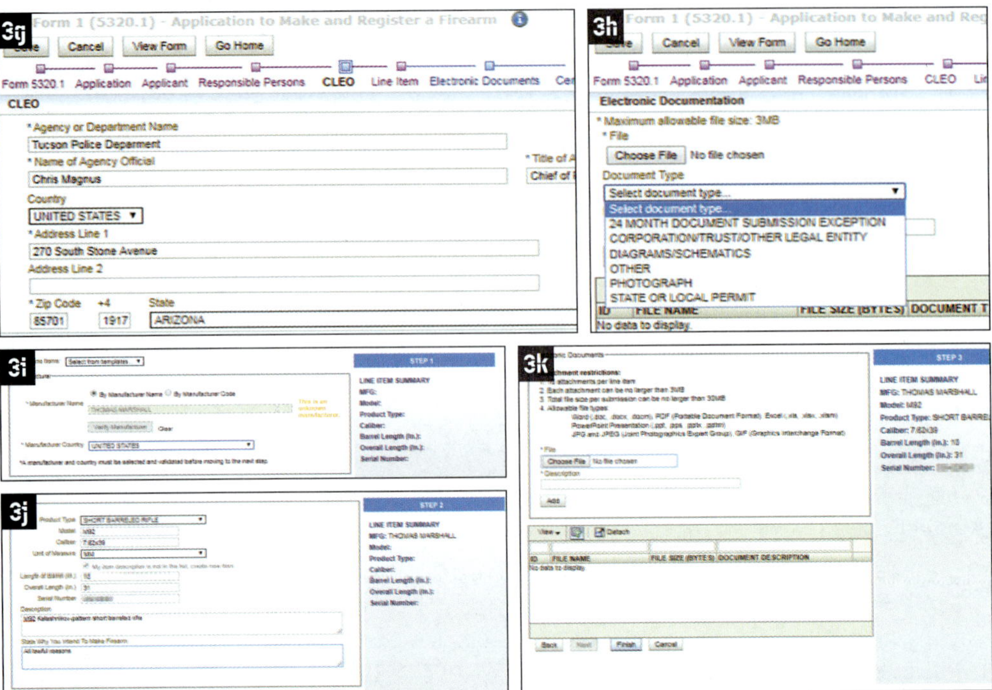

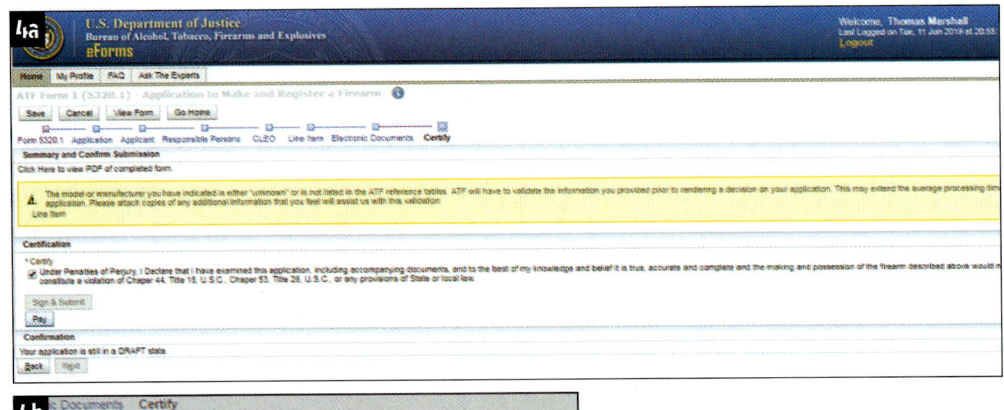

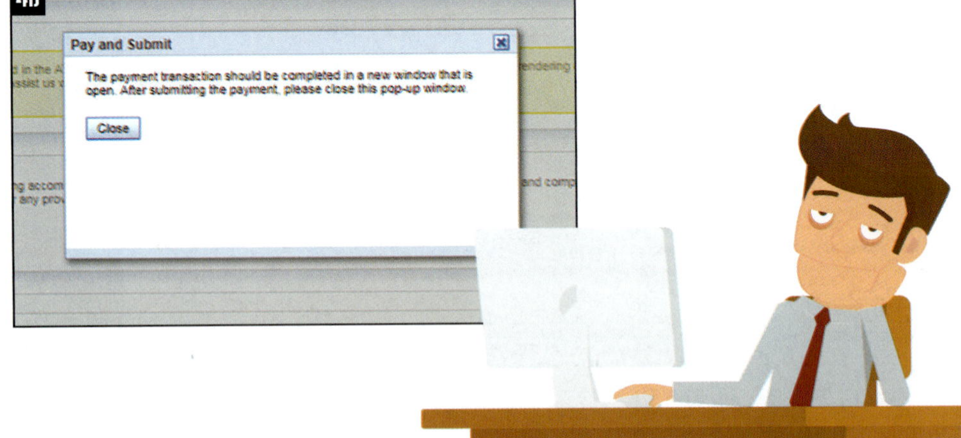

That's it. All in all, the process is a little tedious but, for this first-timer, much less intimidating than initially expected. See the included sidebar for a rundown on engraving requirements. **R**

FRIGGIN' LASER BEAMS

The Internet Was Wrong—You Have to Engrave Form 1 Items

By Dave Merrill

WHY

If you Form 1 an NFA item, disregard whatever purported "truth" a low-information trash weblog told you and listen to what we've heard directly from the BATFE: You have to mark any personally manufactured NFA item. The BATFE has gotten so many questions due to this garbage rumor they've even been stamping some forms reminding folks it needs to take place.

MINIMUM REQUIREMENTS

The depth must be no more shallow than .003 inch with a text size no smaller than 1/16th of an inch. But we actually recommend going deeper and larger. Here's why: If you only engrave to the minimum size and depth, it's incredibly easy to inadvertently cover the marking with paint or something like Cerakote. And so long as you place it someplace inconspicuous, the marking can be more than twice the depth and size without being anything close to obvious.

WHAT

Per 27 CFR § 479.102:

"In the case of a domestically made firearm, the city and State (or recognized abbreviation thereof) where you as the manufacturer maintain your place of business, or where you, as the maker, made the firearm"

There are two ways to go here. If you're making an NFA item from another established item, such as an AR receiver, all you need is the name of the manufacturer and city of manufacturing. However, if you're making a complete NFA item from scratch, in addition to the above you'll need to add a serial number, caliber, and model.

This means if a trust was used to manufacture the firearm, the trust name must be listed, and if you did it as an individual, your name must be listed. Along with the other information required.

Let's say a standard AR is used to manufacture the firearm (as is common), such a marking would look like this:

RECOIL TRUST
PHOENIX, AZ

If an NFA item is built from scratch, such a marking would look like this:

RECOIL TRUST
PHOENIX, AZ
SN: 0006969
CAL: 7.62 MM
MODEL: YEET CANNON

While serial numbers themselves are required to be engraved/stamped/pressed onto metal (à la the metal tab on a Glock), all other information can be on plastic. So don't freak out that you don't have a piece of metal to engrave on with a Glock.

WHERE

While if you're creating a firearm from scratch all makings need to take place in a conspicuous place on the receiver, with an NFA item it's not so straightforward. You're allowed to mark on the receiver, frame, or barrel.

While marking the barrel itself may seem strange, it doesn't when you leave the AR world and start looking at non-ARs like SCARs, AKs, and Thompsons. Note that it says "conspicuous place," and generally that means a place that can be accessed without requiring tools. The easiest place to hide something on an AR is the upper inside of the trigger guard, but you can't put it someplace underneath the pistol grip.

HOW

Our favorite place to get this performed by a large margin is Tarheel State Firearms. They charge $30 for metal receivers with $15 return shipping for any number of firearms sent. However, if you can find someplace local, even at a higher cost, you don't have to bother with shipping. We hit up CNS Engraving in Powell, Ohio, for the photos you see here.

WHEN

The standard route is to send the receiver at the same time as the tax stamp. While it makes a certain sort of sense to only engrave after an NFA item is approved, it's also a helluva lot easier to replace a receiver lost in the mail before it becomes an official Title II item. If there's a place that can perform the service locally, we recommend rolling with one of those, even at an additional cost, to save on time and shipping.

Basically: Roll the dice when you want to do it.

HURRY UP, QUIET DOWN

Silencer Shop and the eForm 4 Process

By Tom Marshall

After years of hinting, teasing, and general government procrastination, we the masses were finally blessed by our ATF overlords with an electronic filing system for Form 4s. The "big deal" is a significantly reduced wait time compared to hand-filled paper equivalents. Since the advent of eForm 1's, wait times for SBRs have dropped down as low as less than two weeks, depending on when you filed. While we've yet to see a Form 4 suppressor stamp turn around that fast, we've seen or heard of 90- to 180-day turnarounds, a vast improvement over the 12- to 15-month average seen previously with paper Form 4s.

Silencer Shop has built their brand not only on selling suppressors, but on making that process as accessible and hassle-free as possible. Once the ATF expanded their eForms infrastructure to include Form 4s, Silencer Shop adapted their own processes to integrate the new technology. We recently went through our first eForm 4 suppressor transfer in order to wrap our heads around how the whole process works.

SET UP YOUR ACCOUNTS

To successfully navigate the process, you must have a user account

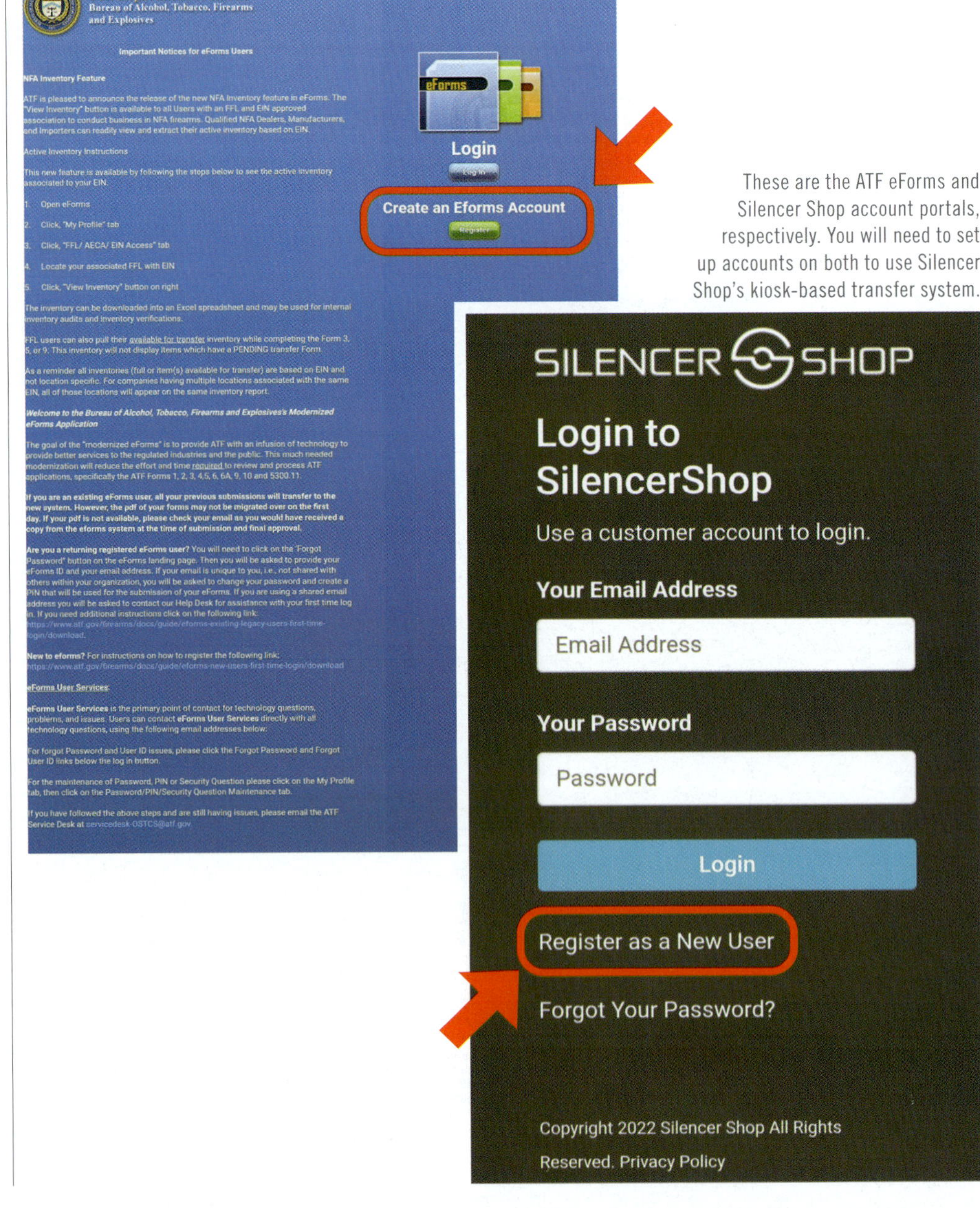

These are the ATF eForms and Silencer Shop account portals, respectively. You will need to set up accounts on both to use Silencer Shop's kiosk-based transfer system.

Above: From the Silencer Shop homepage, click the head-and-shoulder button at the top right to get to the account portal.

set up both with Silencer Shop and the ATF eForms system. For Silencer Shop, visit https://silencershop.com and for ATF eForms, the website is www.eforms.atf.gov. Both websites have buttons to click in order to create a new user account.

This was our biggest learning point. It's absolutely critical that the information you submit to create these accounts matches exactly. First and last name, address, email address, and so on must all be the same, or else you'll run into issues when it comes time to certify your application.

When you create your Silencer Shop account, you'll need to upload a photo of yourself, taken against a white or light-colored background. Don't wear a hat or sunglasses in this photo. The easiest way to do this is to download the Silencer Shop app on your phone. From the app, you can take a selfie on your phone and upload it directly to your Silencer Shop account. This photo will have to updated annually to keep your account current.

Once you create your ATF eForms account, they'll assign you a randomly generated username, but you'll select a password and PIN associated with your account. Save this information, as you'll need it later.

PURCHASE YOUR SUPPRESSOR

There are two ways to purchase your suppressor. One is to buy directly from Silencer Shop's website. Once your web purchase is complete, Silencer Shop will deliver your suppressor to the local partner store of your choice. The website has a searchable listing of brick-and-mortar stores they've partnered with to help you complete your transfer. We used The Hub AZ in Arizona. They're a Powered By Silencer Shop dealer, with an in-store kiosk. Their staff was incredibly helpful guiding us through the process. If you purchase your can from the Silencer Shop website, you can also pay your tax stamp through them at that point.

The other option is to walk into a local partner store and purchase a suppressor from their inventory. Take note: Even if you purchase a

Silencer Shop kiosks are equipped with cameras and scanners to input your photo and fingerprints for NFA transfers.

The Hub AZ, with multiple locations around Arizona, is a Silencer Shop partner dealer.

suppressor from your local store that Silencer Shop doesn't carry, you're still able to use their e-process to complete your transfer. If you purchase the can locally, you'll pay your stamp (plus a processing fee) in person at that store.

Regardless of where you actually purchase from, you'll need to visit your local partner store and use their kiosk to complete your eligibility questionnaire and submit fingerprints. The best part about this process is that the kiosk has a built-in fingerprint scanner, so you won't have to mail hard print cards into the ATF directly.

If you choose to set up a trust for your purchase, you'll be able to do this during the purchase process, either on Silencer Shop's website or kiosk.

CERTIFY AND COMPLETE

Once you've completed your purchase and submitted all required information to your Silencer Shop account, waiting game number one begins. For us, it took about 10 days until we received an email from Silencer Shop saying our order was ready to certify. During the certification process, your local partner store synchronizes the information from your Silencer Shop account with ATF's eForms to begin the background check and stamping process.

If you purchased your suppressor through a trust, you may have the ability to certify your application remotely via video conferencing or phone call. For individual stamps, you'll have to go into your local store and certify in-person on your store's computer. This will require you to enter your ATF eForms username, password, and PIN, and e-sign a couple of validation statements from ATF. This is where we ran into issues, because the information in our Silencer Shop user account didn't match the information in our ATF eForms account. A brief email exchange with Silencer Shop's customer service folks got us squared away on that, and the certification went through without another hitch.

This is when the long wait begins. The average turnaround time we're seeing and hearing about on eForm 4s is six to nine months at time of writing. The stated goal (and our hope) is to get that time down to three months. So far, we've not seen anything that short yet. But once your ATF background check is complete, you'll receive an email from ATF eForms saying your stamp has been issued. You can then download and print out a copy of the stamp from your eForms account. Bring a printed copy of the stamp into your local store to pick up your suppressor, at which time they'll have you fill out a standard 4473 to transfer the item off their inventory book. This will be a no-background-check-needed 4473, since your tax stamp will serve as verification that a check has already been completed.

After that, hit the range and start making not-much-noise with your new suppressor. ■

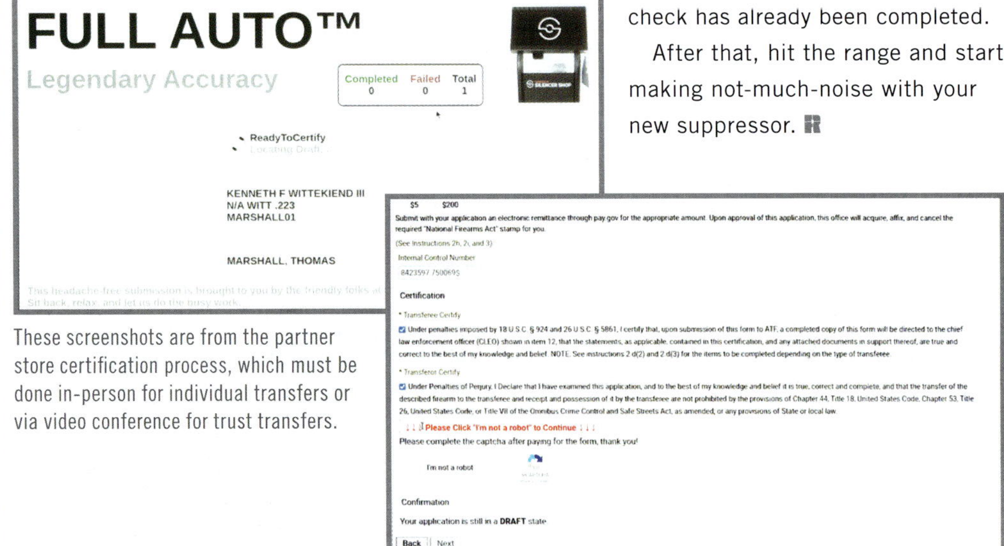

These screenshots are from the partner store certification process, which must be done in-person for individual transfers or via video conference for trust transfers.

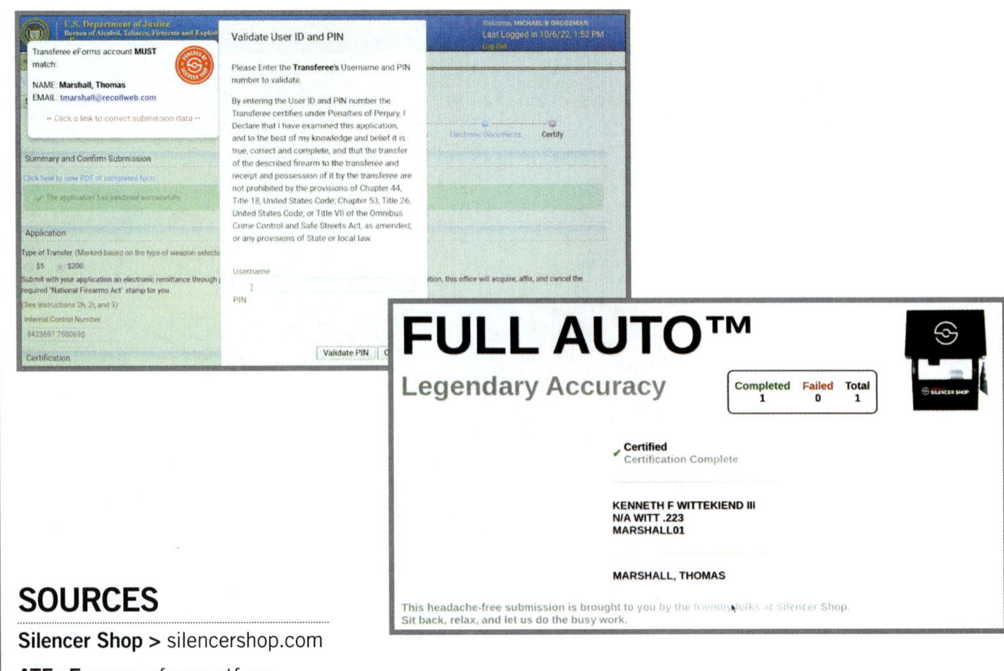

SOURCES

Silencer Shop > silencershop.com

ATF eForms > eforms.atf.gov

The HUB AZ > thehubaz.com

HOW 20TH CENTURY FEAR STAMPED OUT SILENCERS

The "Noiseless Nightmares" of Hiram Maxim

By Kel Whelan

[This is an article excerpt edited to fit this format. Full text available on RecoilWeb]

Today the oft-repeated gun shop lore is that up until 1934 when the National Firearms Act passed, silencers were sold widely via mail-order in the Sears Catalog for $5 or $10 — but their use in poaching during the Great Depression and by gangsters during Prohibition caused manufacturing of the harmless hush-puppies to vanish overnight.

Like so many urban legends, there's half-truths and holes in that narrative. Not to mention the dates don't line up.

Alcohol prohibition, with its unintended consequences of organized — and violently weaponized — crime, didn't take place until 1920. And by that time, many states had already banned silencers. The Prohibition-fueled crime wave wasn't the impetus against them.

Similarly, the Great Depression ran between August 1929 and March 1933. Poaching deer quietly to feed one's starving family would, indeed, make a palatable explanation for why silencers might be under scrutiny. However, Maxim's company had already voluntarily exited the silencer business by the spring of 1930 because the sale or use of their namesake product was banned by many states years earlier.

The NFA most certainly didn't put Maxim out of business overnight; he had already been long chased out of it. All the 1934 NFA law did was merely tie up a two-decade-old anti-silencer period in history with the neat bow of federal homogenization.

BURN THE WITCH

When the silencer was initially invented in 1902, it was enthusiastically admired, then quickly misunderstood and consequentially feared. For a short moment, Maxim's creation was hailed in the common press as a wonder as worthy as those made by Edison, Ford, Tesla, Westinghouse, and the like during that particularly yeasty time in American ingenuity. But newspaper articles turned speedily from admiration to terror. It didn't take long for panicked politicians to envision "the silent killer" would become a widespread worry. And there were, admittedly, a handful of high-

ONE
MAXIM SILENCER
MAXIM
MODEL LM
EXHAUST SILENCER
for
SMALL AIR-COOLED ENGINES ON
LAWN AND GARDEN EQUIPMENT–GENERATORS–PUMPS–COMPRESSORS
INDUSTRIAL MAINTENANCE EQUIPMENT–MARINE APPLICATIONS
MANUFACTURED BY
THE MAXIM SILENCER CO.
HARTFORD, CONN., U.S.A.
THE WOODSMAN
COLT AUTOMATIC
CAL. .22 LONG RIFLE

Hiram Percy Maxim. Photo courtesy Library of Congress.

profile crimes in the early 1900s to touch off the anti-silencer hysteria. The novelty of the few crimes committed while using silencers was enough to fuel the uneducated public's fears.

The first notable case was the horrifying account of Herman Auerbach. Heir to a very wealthy New York candy-manufacturing family, he had sold his share of the business, bet his fortune heavily on real estate investments, and lost big. On the Sunday night of January 31, 1915, his depression finally culminated in his own suicide — and because he was extra awful, the murder of his own wife and children.

While this would be a titillating tragedy for the newspapers to report all by itself, the next morning's *Evening World* newspaper's bold, all-caps headline led with an unusual twist: "USED GUN SILENCER TO MURDER THREE AND KILL HIMSELF." The article reported that Auerbach's second wife and his two "pretty daughters" were killed using a rifle loaded with "dum-dum bullets," adding to the gory contrast of the scene.

The final horror for readers was how his 14-year-old son was left alive to discover the carnage. He was sleeping in a bedroom directly abutting the room where his sister was shot in the head. A Maxim silencer screwed to the end of the .44-caliber magazine-fed rifle used by Auerbach was said to ensure the son stayed asleep while unspeakable murder took place only a wall apart. All this took place in a many-neighbored New York apartment, with no one hearing the slaughter.

But could four suppressed gunshots, the firearm's action, and bullet impact noises be so silent that they weren't even heard through a thin wall a mere foot or so away? The silencer's inventor, Hiram P. Maxim, himself may have had trouble believing it. Regardless, he certainly saw the need to quickly counter the police commissioner's immediate calls for an ordinance that would prevent the sale of silencers.

By February 10, a week after the murders, Maxim ended up coming to New York to demonstrate his silencer in the basement of the police headquarters. The inventor "showed the device to be ineffective on the ordinary revolver and by no means a complete silencer on a rifle." Using the actual rifle Auerbach committed his crimes with, Maxim and his uncle Hudson came to prove to the police commissioner, district attorney, and other officers that the gun's report, although diminished in volume, was unmistakable for anything else than that of a gun.

Newpaper clipping: First Recorded Murder Using Silencer, *New York Times*, Auerbach 1915 "Silent Gun Kills Family of Four."

SILENT GUN KILLS A FAMILY OF FOUR

Herman Auerbach Slays Wife, Daughters, and Himself at 386 Central Park West.

SON, SPARED, FINDS BODIES

Women All Asleep When Rifle with Maxim Device and Dum-dums Ends Their Lives.

FINANCIAL LOSS THE CAUSE

Real Estate Speculator Prepared for Tragedy Several Days Ago, but Gave No Hint of Plans.

Regardless of whether a silencer was actually used or not, the blame continued falling on the firearm, the unusual ammo, and the silencer, rather than on the man who had premeditatedly pulled its trigger. The February 1 edition of the *New York Times* hammered this home with an even more tool-shaming headline that would lead readers to believe that all four persons were victims and only the gun was the antagonist: "SILENT GUN KILLS A FAMILY OF FOUR."

Headlines like this made New York State Senator Irving J. Joseph so aghast that, only two days later, on February 3, 1915, he introduced his bill to make it a felony to sell a silencer in New York.

There were other crimes involving silencers that caught the public's fascination during this era. A year after the Auerbach tragedy, Albany, New York, was terrorized by Harold L. Severy, a young man who had a local machine shop fashion a sleeve gun he concealed in his overcoat. His Stevens firearm also had a Maxim muffler mounted to its muzzle.

Severy's own father claimed, "I think there is no doubt my son is deranged." Six years prior, his son had suffered a head injury during a hazing incident as an art student in Boston. The brain damage resulted in his being sent to two sanitariums. Severy escaped, later surfacing to throw the city of Albany into a week of uneasy fear after he stalked and shot four random persons in the back on a Friday night in January 1916.

Strangely, since they didn't hear a gunshot, each of the stalker's prey continued walking along their way. As they each were shot only once with the diminutive .22-caliber round, it "felt like a sharp sting," but apparently being shot by the low-powered round didn't warrant stopping. The *Advertiser-Journal* of January 29, 1916, wrote, "None of them knew they were wounded until being examined by physicians." The day prior, two women believed they had also been shot at by a youth with an air rifle because they hadn't heard a gunshot either. The entire police department began searching for the would-be assassin, and a well-dressed youth was soon arrested.

Even though, as *The Argus* newspaper reported, Severy was "… possessed of the far-away look, which showed the authorities that his mind was off," Senator Joseph again wasted neither time nor tragedy in pushing his pet plan to criminalize silencers in New York. On February 2, 1916, just a day after Severy's arrest and almost a year to the day since his last bill's attempt to ban silencers, Senator Joseph and Assemblyman John Malone advanced their bill that would make it a felony to sell any

Newspaper clipping: Joseph Bill Introduced ("Legislation Against Silencer").

Legislation Against Silencer.

Senator Joseph, of New York, and Assemblyman John G. Malone yesterday introduced a bill making it a felony to sell a silencer, punishable by imprisonment for five years, except where such contrivances are sold to a military or civic organization. Senator Joseph introduced an identical bill on February 3, 1915, following the use of a Maxim silencer in the killing of a whole family in his district in New York.

Newspaper clipping: "Passes Anti-Silencer Bill Albany, March 8…")

Passes Anti-Silencer Bill.

ALBANY. March 8.—Senator Joseph's bill to prohibit the manufacture, sale, or use of silencing devices on firearms, except in the case of civil or military authorities, was passed in the Senate today.

Page from Seventy-Fourth Legislature of Maine 1909 anti-silencer law.

ACTS AND RESOLVES

OF THE

SEVENTY-FOURTH LEGISLATURE

OF THE

STATE OF MAINE

1909

Published by the Secretary of State, agreeably to Resolves of June 28, 1820, February 18, 1840, and March 16, 1842

AUGUSTA
KENNEBEC JOURNAL PRINT
1909

Chapter 129.

An Act to prohibit the use of Firearms fitted with any device to deaden the sound of explosion.

Be it enacted by the People of the State of Maine, as follows:

Section 1. It shall be unlawful for any person to sell, offer for sale, use or have in his possession, any gun, pistol or other firearm, fitted or contrived with any device for deadening the sound of explosion. Whoever violates any of the provisions of this act shall forfeit such firearm or firearms and the device or silencer, and shall further be subject to a fine not exceeding one hundred dollars, or to imprisonment not exceeding sixty days, or to both fine and imprisonment. Any sheriff, deputy sheriff, constable, inland fish and game warden or deputy inland fish and game warden shall have authority to seize any firearm or firearms and any device or silencer found in possession of any person in violation of this act, and on conviction of the party from whom such firearm or firearms are seized, such firearm or firearms shall be sold, the proceeds to be paid to the state treasurer, and the device or silencer shall be destroyed. Use of firearms fitted with device to deaden sound, prohibited. —penalty. —seizure of firearms, and by whom.

BY-LAWS OF BOARDS OF HEALTH.

Section 2. This act does not apply to military organizations authorized by law to bear arms, or to the national guard in the performance of its duty. Military organizations not affected.

Section 3. In all prosecutions arising under this act, municipal and police courts and trial justices in their respective counties shall have upon complaint original and concurrent jurisdiction with the supreme judicial and superior courts, and all fines, penalties and forfeitures recovered by any person for any violation of this act shall be paid forthwith by the person receiving the same to the state treasurer, to be credited to fines and license fees for the protection of birds and game.

Approved March 24, 1909.

silencer, punishable by five years of imprisonment.

The bill was passed only a month later. It was then signed into law by Governor Whitman on April 6, 1916. This law prevented New Yorkers access to legal firearm-mounted hearing protection — and began cueing up Maxim's exit from the rapidly shrinking, highly restricted market. But, like most laws that attempt to prohibit a behavior by banning an item, it didn't stop human nature.

New York's laws grew more anti-gun with the passage of their Sullivan Act, named for its sponsor, state senator Timothy Sullivan, a notoriously corrupt politician. It made it a felony for otherwise law-abiding New Yorkers to possess or purchase small handguns and other weapons unless they petitioned and paid the government for permission to do so. Human nature being what it is, underground markets sprung up and now possession alone became the criminal act, rather than actual violence.

The initial purchasers of silencers were overwhelmingly law-abiding sportsmen. But that pool of users was now artificially prevented from buying them. Like today's illegal narcotics market has shown, the sort of person who's willing to break one law might also have a proclivity to ignore others. Prior to this date, many people from all walks of life might prefer shooting quietly. But from this point in time, the only news reports written about silencers would of course involve criminal use.

A high-profile example of this occurred in 1924. Eugene Reising, who later went on to firearms-fame as the inventor of the Reising M50 submachine gun, once was arrested for violating the Sullivan Act. His crime was selling some Maxim silencers to the "Cowboy" Tessler Gang, whose members were charged with 81 holdups and at least one murder. Reising served 15 months jail time, but apparently wasn't prevented from returning to the arms trade any more than banning silencers and firearms had stopped the Cowboy Tessler gang from obtaining them.

Like New York, many individual states and cities had banned the silencers outright well before the NFA's ban-by-taxation scheme was hatched. Anti-silencer laws had been passed a decade or more prior in many of the 48 individual states. Some were hunting-based, some required registration, and others simply banned the item's possession, manufacture, or sale without explanation or cause.

In Maine's case, the irony was thick. It was the state of Maxim's father's birth. It was the first state in the U.S. to pass their own alcohol prohibition in 1851. And it was also one of the first states to enact prohibition of the Maxim family's silencers. Maine introduced a law in 1909, the same year Maxim's silencer patent was issued — "An Act to prohibit the use of Firearms fitted with any device to deaden the sound of explosion, Ch. 129, § 1."

One could make a case that the impetus of the Maine law wasn't any concern of violent crime. Chapter 130 of their 1909 law says that, "Any sheriff, deputy sheriff, constable, inland fish and game warden or deputy inland fish and game warden shall have authority to seize any firearm or firearms and any device or silencer found in possession of any person ... and all fines, penalties and forfeitures recovered by any person for any violation of this act shall be paid forthwith by the person receiving the same to the state treasurer, to be credited to fines and license fees for the protection of birds and game."

Similarly, Wyoming's first 1921 anti-silencer law is an afterthought buried behind 58 pages of wildlife codes mandating mundane things such as the size of red bandanas to be worn in hunting fields, how deer tags are issued, and regulations on the use of hunting dogs. It seems there was no worry there about violent crime, but rather concerns over what hunting firearms were deemed "sporting" enough.

North Carolina's 1925 law was solidly based on hunting concerns, even calling out Hiram's brand by name. "It shall be unlawful to trap for bear or to run or hunt deer with dogs or to use while hunting any gun having a 'Maxim silencer' or any other device thereon that will muffle the

Newspaper clipping: "BARS MAXIM SILENT GUN" (PITTSBURG BAN from *New York Times*). The penalty for possession of a muffler in Pennsylvania was to be greater than possession of an actual firearm.

BARS MAXIM SILENT GUN.

Pittsburg Police Search for Users of Invention—"Aid to Crime."

Special to The New York Times.

PITTSBURG, Penn., March 7.—Users of the Maxim gun silencer had better keep away from Pittsburg. The "silencer" is not to be permitted to do business here, the Police Department has already taken action, and persons found in possession of a gun with the new device attached will be sent as far as the law will permit them to be sent.

The matter was brought to the notice of the Police Department by a citizen, who pointed out the almost unlimited opportunities for crime presented in the use of the "silencer."

Superintendent of Police Thomas A. McQuaide said to-night: "There is no question in my mind but that the use of the 'silencer' will prove disastrous to the peace of every city where precautions against its use are not taken. With a 'silencer' attached to his revolver a thug or murderer could stand 100 feet away from his victim, shoot him, and then make his escape without fear of detection.

"Highwaymen at present seldom use a revolver except to intimidate pedestrians. The risk of shooting is too great; the discharge of the weapon makes too much noise and attracts too many people. But with the silencer in use this would be different."

Mr. McQuaide says his men have been instructed to be vigilant in searching out users of the new contrivance. The penalty imposed by the Pennsylvania law is up to a year's imprisonment and a heavy fine. At present persons caught carrying concealed weapons are usually disposed of with a short sentence to jail or the workhouse.

report of such gun, nor shall any gun be used that does not produce when discharged the usual and ordinary report."

Following Maine, there were anti-silencer and anti-silencer hunting laws put on the books in Vermont (1912), Minnesota (1913), Michigan (1913), New York (1916), New Jersey (1920), Wyoming (1921), Pennsylvania (1923), North Carolina (1925), Massachusetts (1926), Rhode Island (1927), California (1933), and other jurisdictions and cities.

While some laws were clearly rooted in trying to hold back modern advancements in hunting technology, some of these states' early anti-silencer laws exempted police and military. Clearly that aspect wasn't included for hunting. The silencer was introduced by Maxim for sporting purposes, but military use was certainly also envisioned. The states that included language exempting government use clearly feared crime and warfare.

Pittsburg's 1909 decry against silencers has some of the earliest recorded police backing. It also shows there was literally nothing other than fear motivating the law. "The matter was brought to the notice of the police department by a citizen, who pointed out the almost unlimited opportunities for crime presented in the use of the 'silencer.'" While the writings of the time admitted that handguns are seldom used by highwaymen robbers, it was hypothesized that criminals would be tempted to use their firearms more often if they were quieter. This law wasn't based on any evidence, only one citizen's worry of what "could be."

The words of a police superintendent from a major metropolis must carry some weight; in a press conference, one man merely opining that silencers "will prove disastrous to the peace of every city where precautions against its use are not taken" was apparently enough "proof" to get laws on the books. Pittsburg's policy was based entirely on conjecture. The penalty for possession of a muffler in Pennsylvania was to be greater than possession of an actual firearm.

Vermont's 1912 law clearly spelled out that their goal was to "Prevent the Manufacture, Sale, or Use of Gun Silencers." It was clearly of the nature that envisioned violent crime, as it exempted military purposes, and contained no language concerning hunting.

Massachusetts' 1926 law also follows in those prejudicial footsteps.

Massachusetts legislative record page: "HOUSE.... No. 38" (Boston Police Massachusetts anti-silencer recommendations).

HOUSE . . . . . No. 38

The Commonwealth of Massachusetts

City of Boston, Police Department,
Office of the Commissioner, Dec. 1, 1925.

To the General Court of the Commonwealth of Massachusetts.

In compliance with section 33 of chapter 30 of the General Laws, as amended by section 43 of chapter 362 of the Acts of 1923, I have the honor to submit herewith such parts of the annual report of the Police Commissioner for the City of Boston for 1925 (Pub. Doc. No. 49) as contain recommendations for legislative action, with accompanying bills.

Yours respectfully,

HERBERT A. WILSON,
Police Commissioner for the City of Boston.

2 HOUSE — No. 38. [Jan.

RECOMMENDATIONS.

1. Relative to Firearm Silencers.

I recommend that further legislation be enacted to prevent the sale or use of silencers or any instrument, attachment, weapon or appliance for causing the firing of a gun, revolver, pistol or other firearm to be silent or intended to lessen or muffle the noise of the firing of the same. Such devices are now being manufactured and placed on sale. This instrument has recently been used in other cities outside this Commonwealth and legislation in this direction is necessary to assist the police in apprehending offenders who use this device in the commission of crime.

2. Prohibiting the Sale or Distribution of Magazines, Newspapers and Periodicals containing Advertisements of Firearms.

I again recommend for consideration such legislation as will forbid the sale of magazines or periodicals advertising the sale of firearms within the Commonwealth. If such legislation was enacted into law, it would in my opinion help to stop the indiscriminate distribution of firearms by mail order houses, many of such firearms now finding their way into the hands of youths and other irresponsible people.

While I agree that such legislation would be more effective if passed by the Congress of the United States, yet until this is done, I believe that this Commonwealth should lead the way and do all possible to curtail such sales. If laws can be enacted to prohibit the sale of magazines containing obscene pictures and stories not fit for publication, they can likewise be passed to stop the advertising of these death dealing weapons.

Boston's Police Commissioner, Herbert A. Wilson, lobbied for legislative action to "prevent the sale or use of silencers." He mentioned the devices had "recently been used in other cities" and that "legislation is necessary to assist the police in apprehending offenders who use this device in the commission of crime." Specifically mentioning use outside the Commonwealth, and having no valid examples that occurred in Massachusetts itself, he may have been referring to the two New York incidents of over a decade prior.

HUNTERS, GAME WARDENS, AND RACISTS: UNITED AGAINST SILENCERS

The sportsman's community of the day was divided in debate over this new invention. Just like today, hunters of the 1900s comprised a complex, diverse group of individuals. Some were content to wipe out the bison and other species wholesale, while others were careful conservationists. Silencers were lumped into a class of "unsportsmanlike" products by Luddite, hyper-conservative hunters.

The media again played a part in advancing the negative narrative. An opinion piece published in the 1909 *New York Times* illustrates some hunters and hunting organizations of the day's view on the controversial cans. "Who needs a silencer? The Black Hand (an immigrant extortion and kidnapping gang) assassins, but surely not a sportsman!"

William T. Hornaday, an influential conservationist who served as the director of the Bronx Zoo and once actually put a Congolese black man on display in a cage, was enthusiastic about banning modern firearms of all sorts — often as a hand-in-hand move to advance his openly racist views.

Hornaday argued that all pump-action guns should be banned, as should all semi-automatics and all "silencers." In his 1913 book *Our Vanishing Wildlife: Its Extermination and Preservation*, he blamed the decimation of wildlife on not only automobiles, but black people, immigrants, and the poor.

He urged new laws to "prohibit the use of firearms by any naturalized alien from southern Europe until after a 10 years' residence in America." Wildlife was vanishing because "the Italians are spreading, spreading, spreading. If you are without them to-day, to-morrow they will be around you. Meet them at the threshold with drastic laws, thoroughly enforced." One such "drastic law" he may have influenced was the National Firearms Act of 1934.

President Theodore Roosevelt was clearly on the progressive side of technology and science. Here, he writes to show his gratefulness and excitement over having his rifle barrels threaded for silencers in a personal correspondence to Hiram P. Maxim.

The Outlook
287 Fourth Avenue
New York

Office of
Theodore Roosevelt

March 12th 1909.

My dear Mr Maxim:

The 30-30 Winchester has come; I shall try it at once, and I need hardly say how much obliged I am to you. General Crozier has not yet forwarded the other two guns to the Winchester Company, but he will doubtless do so in a reasonable time. I shall look forward to trying them in Africa.

With hearty thanks for your courtesy,

Sincerely yours,

Theodore Roosevelt

Mr H. P. Maxim,

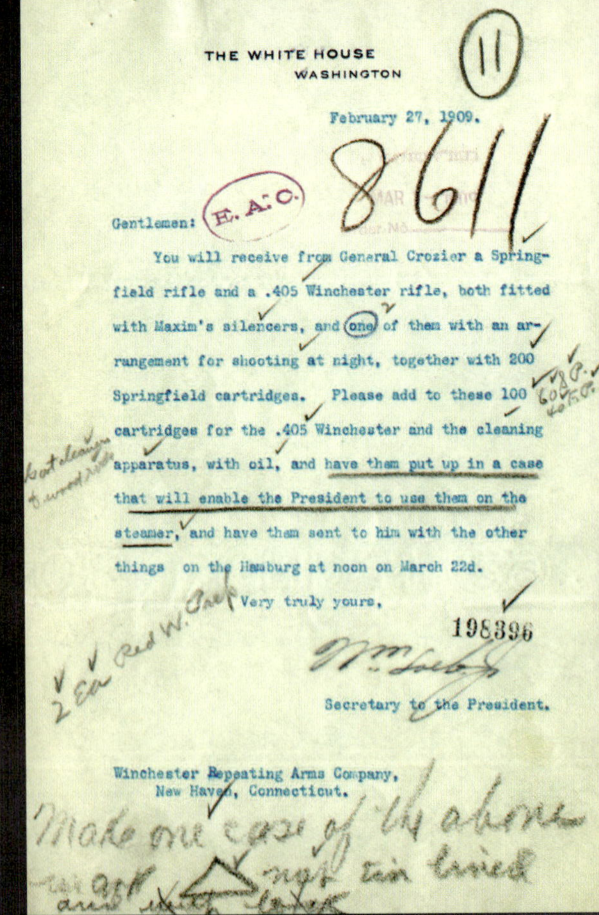
THE WHITE HOUSE
WASHINGTON

February 27, 1909.

8611

Gentlemen:

You will receive from General Crozier a Springfield rifle and a .405 Winchester rifle, both fitted with Maxim's silencers, and one of them with an arrangement for shooting at night, together with 200 Springfield cartridges. Please add to these 100 cartridges for the .405 Winchester and the cleaning apparatus, with oil, and have them put up in a case that will enable the President to use them on the steamer, and have them sent to him with the other things on the Hamburg at noon on March 22d.

Very truly yours,

198396

Wm. Loeb

Secretary to the President.

Winchester Repeating Arms Company,
New Haven, Connecticut.

Make one case of the above

THE "SILENCER" FOR FIREARMS.

Who shall be authorized to carry and use firearms equipped with Mr. HUDSON MAXIM'S "silencer"? No private person, surely. A true sportsman would not use it, and the "pot-hunter" must be forbidden to hunt silently. The burglar, the highway robber, and the Black Hand assassin are the only other persons to whom it could be of advantage. The Scientific American says this week that through the "silencer" the fatal bullet could be sped to its mark "on a crowded thoroughfare and in broad daylight, without there being any evidence of smoke or report to show whence it came."

We, therefore, see a reason for the passage of Assemblyman JOSEPH'S bill making it a felony, punishable by imprisonment for not more than five years, to make or sell the instrument except for duly authorized military or civic organizations, or to have or to carry it concealed upon the person. The bill refers not merely to the private fitting of the "silencer" to rifles, but to "any gun, revolver, pistol, or other firearms." The objection that revolvers and pistols cannot be successfully muffled will be valid only until some slight improvement in the invention shall adapt it to their use. The prohibitions of the bill cannot in this respect be too broad and explicit.

SPORTSMAN'S SILENT GUN

Would Exterminate Game Most Effectually of Recent Inventions.

To the Editor of The New York Times:

If, as Mr. Hiram Maxim claims, his silencer for gun fire would be of no assistance to a conscienceless pothunter in the extermination of game "worth protecting," as stated in your editorial of to-day, of what advantage does he claim it will be to a game hunter, and why is he so anxious to be allowed to sell it to them? Surely, the elimination of sound and recoil would hardly compensate one not expecting bigger kills for using the awkward attachment.

As for its use for target practice only, there could be no objection to it, as in that capacity it could do no harm, but every true sportsman who has the protection of our game at heart will agree with Dr. Hornaday that the use of the silencer would contribute more to the extermination of game than has any invention of recent times, not excluding the automatic gun.

Mr. Maxim's arguments (?) in favor of its use for sporting purposes can hardly be viewed as disinterested from a sportsman's standpoint, and, when sifted down, resolve themselves into the assumption that he "needs the money." As an inventor of arguments Mr. Maxim is miles behind himself in other fields. The crux of the matter is that by allowing the use of the silencer for game hunting Mr. Maxim and the pothunters will be the ones benefited, but if its use is prohibited by law all decent sportsmen will be benefited. It should not be difficult for our legislators to decide between these two propositions. W. K. B.

New York, April 27, 1909.

SATURDAY, JUNE 19

WATCH USE OF SILENCER

NEW NOISELESS METHOD OF FIRING GUNS MEANS MORE WORK FOR THE STATE GAME WARDENS.

The new Maxim gun silencer threatens to invade the field of sports, and game wardens are trying to keep cases on the sale of the contrivance that reduces the report of a firearm to a mere puff. Use of the silencer would make the detection of illegal hunting more difficult than at present, and there is plenty of work for the warden under present conditions. It would also give an unfair advantage over the game, as there would be no report to warn survivors of the first fire.

Game Warden Uhlig of Spokane county says that none of the silencers has been sold in this city, to his knowledge. He, with other wardens of the state, is keeping what surveillance he can on stores that deal in firearms. The wardens cannot prevent sale of the silencer, but they can keep a record of names and addresses of purchasers.

In a number of states there is a penalty for using the silencer for unlawful purpose and the wardens take the view that the only lawful purpose of the Maxim invention is use in warfare. Genuine sportsmen and periodicals devoted to hunting are pushing a campaign against the silencer.

Some opinion articles from the era. Newspapers from Oregon to New York spread the fear against silencers. Even back then, the press got it wrong: Hiram's uncle Hudson is blamed for the silencer. He may have invented smokeless gunpowder, but it's Hiram's silencer. Facts are hard!

TRUTH FELL ON DEAF EARS

It wasn't until the late 1800s that it was even known that exposure to high sound levels could lead to permanent hearing loss. Earplugs weren't patented until 1864, and disposable earplugs in 1914. Hearing protectors weren't even generally available until the 1920s. Even then, these devices were marketed to help people cope with the high sound levels of the "big city," and not known to protect against permanent hearing loss. A medical journal study explains the zeitgeist: "The pervasive attitude of the early 1900s was that hearing loss could be prevented by developing a tolerance to noise. Consequently, any attempts to avoid loud sounds or to protect oneself from them were interpreted as weakness."

One hundred years ago, hearing protection was barely thought of, much less used. Men went deaf doing manly things, earplugs were for sissies, and that was just the way life in the early 1900s was viewed. Fears of crime — whether real or imagined — were informed and flavored by only a handful of actual incidents, and more by xenophobia and racism. In the days of the Second Industrial Revolution, rapid advancements in science and technology that could be envisioned to potentially assist criminals were met with ignorant, quickly passed laws.

Today, we know that firearm-mounted hearing protection benefits public health far more than it enables bad actors. Recent studies of federal prosecutions show that very few silencer "crimes" have been processed, and of those, "more than 80 percent of federal silencer charges are for non-violent, victimless crimes." Dave Kopel writes, "In other words, the possessor was not allowed to possess firearms, ammunition or silencers ... but was not misusing the item. In only 2 percent of the cases was the firearm discharged. [A]nalysis of California cases from 2000-2005 found similar results. Thus, the misuse rate for lawfully purchased suppressors appears to be very low."

Most all of the U.S. states have ditched their antiquated anti-silencer laws as the health benefits of silencers are continually proven to outweigh fears of criminal use. At this point, the outdated 1934 federal law stands almost alone in prohibiting a simple, century-old muffler.

AR-15
RECOIL Magazine's Collection of Unique and Exceptional ARs
COMPILED BY THE EDITORS OF RECOIL MAGAZINE

DIY GUNS
RECOIL Magazine's Guide to Homebuilt Suppressors, 80 Percent Lowers, Rifle Mods
And More!
COMPILED BY THE EDITORS OF RECOIL MAGAZINE

CCW
RECOIL Magazine's Guide to Concealed Carry Training, Skills and Drills
COMPILED BY THE EDITORS OF RECOIL MAGAZINE